Machining of Laminated Polymer Nanocomposites

nova science publishers
www.novapublishers.com

Manufacturing Technology Research

More information about this series can be found at
https://novapublishers.com/product-category/series/manufacturing-technology-research/

Polymer Science and Technology

More information about this series can be found at
https://novapublishers.com/product-category/series/polymer-science-and-technology-series/

Jogendra Kumar
Rajesh Kumar Verma
Yadvendra Kumar Mishra
Rajneesh Kumar Singh

Machining of Laminated Polymer Nanocomposites

nova
science publishers

www.novapublishers.com

DOI: https://doi.org/10.52305/NWIJ8428

Library of Congress Cataloging-in-Publication Data

ISBN: 979-8-89530-733-5 (Hardcover)
ISBN: 979-8-89530-805-9 (eBook)

Published by Nova Science Publishers, Inc. † New York

Contents

Preface

In recent years, researchers working in material science have targeted composites due to their ability to tailor properties according to end application. The laminate polymer nanocomposites are novel engineering materials with nano-scale reinforcement. It has become a suitable alternative to conventional materials for lightweight and structural applications. Nanomaterials have emerged as the most promising nanofiller for polymer nanocomposites due to their exceptional mechanical and electrical properties. The carbon nanofiber-reinforced epoxy composites possess superior electrical, mechanical, and thermal properties. The wide acceptance of various engineering applications brings together researchers from academia and industry to study the machining behavior of hybrid materials, as well as the predictive modeling and optimization of machining parameters. The machining of polymer composites is significantly different from conventional metal machining. The composite system consists of two distinct phases, i.e., matrix and reinforcement, which come under the cutting tool during machining. The machining behavior of fiber-reinforced polymer nanocomposites largely depends on the nature of support, the amount of reinforcement within the matrix, and the material behavior. Various types of defects are associated with the machine component during the machining. An adequate knowledge of and understanding of the machining behavior is necessary. This is needed to identify the most favorable environments. Therefore, this book provides systematic information on proposed polymer laminated nanocomposites initiating from the development phase, mechanical characterization, machining, machinability aspect, parametric appraisal optimization, and predictive modeling for better knowledge. The results of this book will help with the exploration of fiber polymer nanocomposites for product applications.

Acknowledgments

This book was made possible thanks to the help, support, and guidance of several prominent individuals and scholars. First, my most profound gratitude goes to the Almighty, who endowed me with health, grace, wisdom, patience, and direction to complete this work. My most profound appreciation goes to the SCSP DST, TIDF-IIT Guwahati (Project-1019) for their continuous support and direction in writing this book.

I am incredibly thankful to the Chancellor and H. E. Governor of Uttar Pradesh and the Government of Uttar Pradesh state for providing us with infrastructure and, other technical support for completing this task. I want to express special thanks to Professor Samsher, Honorable Vice Chancellor, Harcourt Butler Technical University Kanpur, for creating a research-oriented environment in the university and continuous direction for the timely completion of this assignment. Thanks to his exemplary visionary leadership and infrastructural facility support, this work was completed on time. I am extremely grateful to him for motivating me to think independently and good-naturedly and for solving numerous problems with perseverance.

We are highly appreciative and grateful to the Advanced Polymer Composites Laboratory of Harcourt Butler Technical University, Indian Institute of Technology Kanpur, Central Institute of Petrochemicals Engineering and Technology (CIPET) Lucknow, and Babasaheb Bhimrao Ambedkar University (BBAU) Lucknow for their constant laboratory support in completing this work. We are very grateful for the continuous support of Dr. Shivi Kesarwani, Dr. Kuldeep Kumar from MMM University of Technology Gorakhpur, and Dr. Devendra Kumar Singh (HBTU Kanpur). This work is impossible without the blessings and kind support of my global research collaborators from various reputed institutes.

The authors are very grateful to the reviewers, editorial members, and NOVA publication staff for their valuable time and effort on this book. Their efforts were significant in producing this book in its present form. We could not have accomplished this milestone without their constant and consistent advice, support, and collaboration.

Jogendra Kumar

List of Abbreviations

0D	Zero Dimension
1D	One Dimension
2D	Two Dimension
3D	Three Dimension
Adj MS	Adjusted mean square
Adj SS	Adjusted sum of squares
AFM	Atomic Force Microscopy
AHP	Analytic Hierarchy Process
Al_2O_3	Aluminium Oxide
ANN	Artificial Neural Network
ANOVA	Analysis of Variance
AWJM	Abrasive Water Jet Machining
BBD	Box Behnken Design
CCD	Central Composite Design
CED	Cause and Effect Diagram
CF/GF	Carbon Fiber/Glass Fiber
CFRP	Carbon Fiber Reinforcement Polymer
CMC	Ceramic Matrix Composite
CNC	Computerised Numerical Machine
CNMs	Carbon Nano Materials
CNO	Carbon Nano Onion
CNT	Carbon Nanotube
CNT/CF	Carbon Nanotube/Carbon Fiber
COCOSO-PCA	Combine Compromise Solution-Principal Component Analysis
Cu	Copper
CVD	Chemical Vapor Deposition
DF	Degree freedom
DFA	Desirability Function Analysis
DM	Decision Making

DMA	Dynamic Mechanical Analysis
EDM	Electric Discharge Machining
EMI	Electromagnetic Interference
EPD	Electrophoretic Deposition
ESD	Electrostatic Discharge
FEM	Finite Element Method
FG/PU	Functionalized Graphene /Polyurethane
FIS	Fuzzy Inference System
FRP	Fiber Reinforcement Polymer
FTIR	Fourier-Transform Infrared Spectroscopy
F-value	Fisher value
G/CFRP	Graphene/Carbon Fiber Reinforcement Polymer
GA	Genetic Algorithm
GFRP	Glass Fiber Reinforced Polymer
GFs	Glass Fibers
GNP	Graphene Nano Palate
GNP/CTBN	Graphene Nanoplates/Carboxyl Terminated Butadiene Acrylonitrile
GNP/MWCNT	Graphene Nanoplate/Multiwall Carbon Nanotube
GNP/short CFs	Graphene Nanoplate/Short Carbon Fibers
GNRs	Graphene Nanorods
GNWs	Graphene Nano walls
GO Foam/CFs	Graphene Oxide/Short Carbon Fibers
GO	Graphene Oxide
GO/ATP	Graphene Oxide/Attapulgite Nanorod
GO/CFRP	Graphene Oxide/Carbon Fiber Reinforcement Polymer
GRA	Grey Relation Analysis
GRA-PCA	Grey Relation Analysis-Principal Component Analysis
GRC	Grey Relation Coefficient
GRG	Grey Relation Grade
H	Hardener
HB	Higher is Better
HDT	Heat Deflection Temperature
HSS	High Speed Steel
IFSS	Interlaminar Flexural Shear Strength
ILSS	Interlaminar Shear Strength
LB	Lower is Better

MMC	Metal Matrix Composite
MOORA	Multi-Objective Optimization on the Basis of Ratio Analysis
MRR	Material Removal Rate
MWCNT	Multi-Wall Carbon Nanotube
Ni	Nickel
NR	Nanorod
NT	Nanotube
OA	Orthogonal Array
PCA	Principal Component Analysis
PCs	Principle Components
pGO/CF	Pristine Graphene Oxide/Carbon Fiber
PMC	Polymer Matrix Composite
PSO	Particle Simulating Optimization
Pt	Platinum
PVA	Polyvinyl Alcohol
P-value	Probability value
R	Resin
rGO	Reduced Graphene Oxide
RSM	Response Surface Methodology
Ru	Ruthenium
SA	Simulating Annealing
SD	Standard Deviation
SEM	Spectroscopy Emission Method
Seq SS	Sum of squares
sGO/CF	Salinized Graphene Oxide/Carbon Fiber
SiC	Silicon Carbide
SiO_2	Silicon Oxide
SR	Surface Roughness
SWCNT	Single-Wall Carbon Nanotube
TGA	Thermogravimetric Analysis
TiAlN	Titanium-Aluminium Nitride
TOPSIS	Technique for Order of Preference by Similarity to Ideal Solution
VARTM	Vacuum Assisted Resin Transfer Molding
WPCA	Weightage Principal Component Analysis
XRD	X-Ray Diffraction

Chapter 1

Introduction to Laminated Polymer Nanocomposites and Machining Interfaces

1.1. Introduction to Materials

Materials are crucial to the development of various types of components and structures in our society. The Stone, Bronze, and Iron Ages saw material progress in humanity (the industrial revolution). After World War II, industrial chemistry generated many synthetic polymeric materials, resulting in the Polymer Age. Polymer science studies deal with the properties of polymers and polymer-based composite materials. In combination with low density, polymer composite materials have excellent strength and stiffness. Higher polymer materials are suitable for economic demands to save weight in several product (Mohammed et al., 2015; Seydibeyoğlu et al., 2020). The composite materials in polymers are a relatively new category of materials and are thus not used as metallic materials yet. As the composites still consist of a combination of various kinds of macro-level materials, they cannot be considered traditional engineering materials having homogeneous structures.

Figure 1.1 depicts the challenges in locating suitable material for application (Sun & Zhao, 2020). The ASHBY chart illustrates the plot of material selection criteria based on strength vs density. It suggests that metals have strength, but due to their increased density, they are not particularly ideal material for use in structural applications. Laminate polymer composites (LPC) are stronger and lighter than traditional materials. The integration of nanoparticles into LPC, which has a synergistic effect, has significantly enhanced their desired properties (Ashby et al., 2004; Findik & Turan, 2012). In polymer nanocomposites, nanomaterials like CNT, CNO, GNP, GO, rGO, SiC, and Al2O3 are widely used (Kato et al., 2012). Previous research showed that carbon-based nanomaterials (CNMs) such GO, rGO, CNT, MWCNT, and others have a wide range of characteristics and greater possibilities in polymer (Ray & Easteal, 2007). Research has shown that graphene nanoparticles are hundreds of times stronger than steel (Cao et al., 2020; C. Lee et al., 2008). Carbon fibre reduced the weight of its components by 60% in comparison to steel and by 30% compared to aluminium (Ashby, 2013; Ashby et al., 2004).

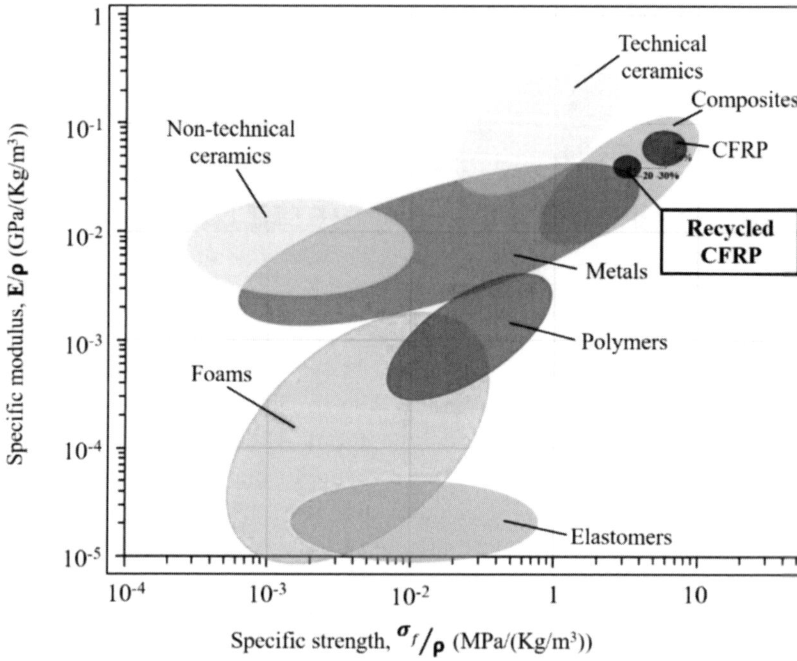

Figure 1.1. Material selection chart (ASHBY chart) (Ballout et al., 2022).

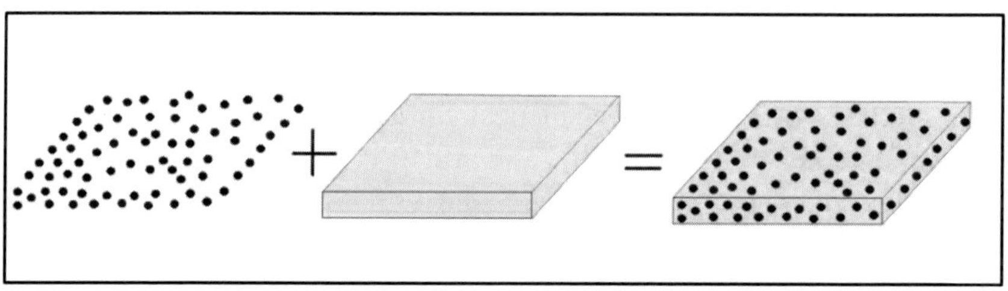

Figure 1.2. Introduction of composite.

1.2. Introduction to Composites

A binding agent or matrix is one of the components of composites that binds the reinforcement in position (shape/size) (Mohammed et al., 2015). The properties and behavior of the composite material are influenced by the retention of both the filler and matrix identities while the filler and matrix are combined. The structure of the constituents is combined with their respective functional phases without the formation of a chemical compound

(Horoschenkoff & Christner, 2012). In recent decades, academics and researchers have been looking deep into the composites research field. Composites consist of low constituents and move to a high-strength element as shown in Figure 1.2, which depicts the composite material phase.

1.3. Classification of Composite Materials

In this section, a broad classification level and its significance have been described in a detailed manner. Generally, composite material is classed into two subsections: composite based on matrix and composite based on reinforcement.

1.3.1. Classification Based on Matrix Type

The matrix can usually categorize composite materials in three sections. These are composite in Metal Matrix (MM), Ceramic Matrix (CM), and Polymer Matrix (PM) (Gay et al., 2002). Figure 1.3 displays the matrix-based classification of composite materials. Composites in the metal matrix have several benefits over monolithic metals, such as a higher module, higher specific strength, improved temperature properties, and less thermal stability. Because of its high-strength properties, MM composite is being considered for a wide range of applications, including compressor blade nozzles, steam turbines, and load bearing (Basu & Debnath, 2015; Singh, 2014). One of the primary goals in the manufacture of ceramic matrix composites is to improve durability. Usually, it is intended, and indeed often discovered, that ceramic matrix composites would increase in strength and stiffness at the same time (Jannatyha et al., 2020; Koch et al., 2006).

With their improved features, LPC plays a crucial role in the production sector in the twenty-first century (Cui et al., 2019). A new generation of composites is created by synthesizing novel lightweight composites with superior mechanical properties. Polymer (epoxy) based composites are frequently utilized in this series because of their higher degree of cross-linking and high aspect ratio. The use of polymer composites in manufacturing has been widely acknowledged in the last few years. The polymers play a primary role in composite families due to their desired properties. The aviation, marine, aerospace, and automotive industries heavily utilise polymer composites because of their improved mechanical qualities and extended fatigue life (Rubino et al., 2020). It can effectively replace more common technical materials like metal, alloys and plastics. The chosen building blocks must exhibit exceptional resistance to deterioration and novel physical and structural features, including a low point-specific weight (Brebu, 2020; Uygunoglu et al., 2015). It is capable of ensuring both safety and economic performance in these multifunctional circumstances. In composites, the matrix material is the main phase that envelops and provides the polymer with its characteristic structure. Its cross-link connections and ability to support loads are both dependent on the polymer matrix phase. The aviation industry and other industries require lightweight components and demand their products to have low weight and versatility. Polymers can be separated into thermoplastic composites and thermoset composites based on the properties of the polymer matrix employed, as depicted in Figure 1.3. Thermoplastics are temperature-sensitive since they can liquidate at temperatures above their melting point during the composite manufacturing process. The individual molecules in thermoplastics are not bonded chemically. On the other hand, Van der Waals' secondary bonds are formed by weak intermolecular forces (Mohammed et al., 2015). Lack of a hard binding force, and adequate heat energy can be used to interrupt the thermoplastic materials. With the addition of heat, the molecules' tendency to move relative to each other allows for further transition and reconfiguring. The material solidifies when the heat source is removed; therefore, it may be reconfigured or recycled with the required application of pressure and heat. Thermoplastic products seem to be more ductile than other composites because of these distinguishing features. There is a realistic limit to the amount of thermoplastic reprocessing due to rising oxidation in the molten state. Subsequently, the liquid resin solidifies, and thermoset materials undergo a chemical reaction during the manufacturing process. In this condition, crosslinking chemically bonded to the molecules together, resulting in a rigid, three-dimensional network framework and an irrevocable composite. Smaller unlinked molecules known as monomers are found in the resin material, which is made of carbon. The chemical reaction is accelerated by applying a curing agent, also known as a hardener, which is nothing more than a catalyst. Oxidized polymers, polyvinyl, polyurethane, and phenolic are widely used in thermoset matrices.

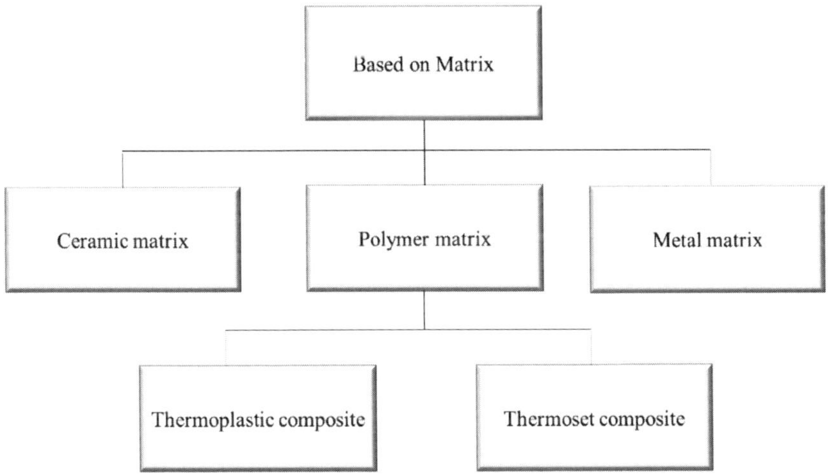

Figure 1.3. Composite classification based on matrix.

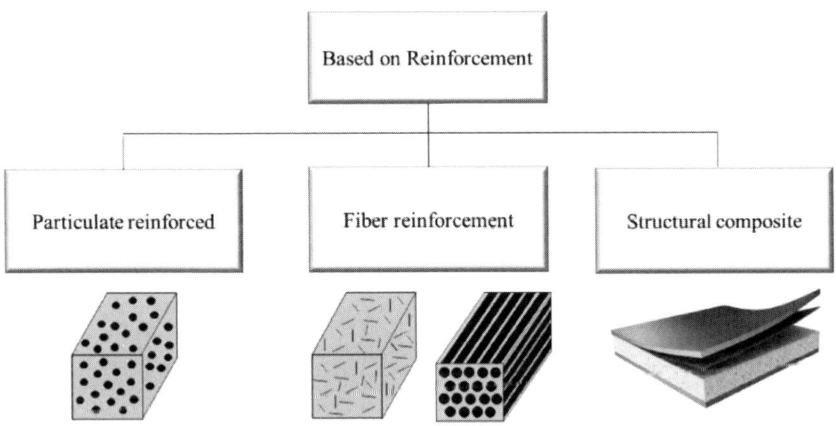

Figure 1.4. Classification of composite based on reinforcement.

1.3.2. Classification Based on Reinforcement Type

The structure constituent determines the intrinsic configuration of the composite. It links the dispersion and matrix phases in the form of bonding. Based on the reinforcement, composite can be categorized into three groups. (a) Particulate Composites, (b) Fiber Reinforced Composite, and (c) Structural Composite. The composite particles are in different forms and dimensions and are randomly distributed in the matrix. Short-fiber composites are the discontinuous type used in composite reinforcement materials. The diameter of the fiber here is minimal in contrast with the length of the reinforcement. The fibers can be distributed uniformly to create a one-way structure. Continuous fibers are incorporated into the composite and bear uniaxial strength, resulting in increased material strength. Figure 1.4 shows the reinforcement-based categorization of composite materials.

1.4. Reinforcement Materials

Polymers have been widely used in the manufacturing sector over the past three decades to satisfy consumers' demands and meet altering market needs. Composites made from polymers and modified polymers provide a number of benefits. They distinguish themselves from traditional production materials by providing several distinctive qualities, such as low density, increased strength, simplicity of processes, economic viability, etc. Despite this, thermoplastic composite materials in many engineering applications were unable to meet the mechanical properties that customers were looking for. By including various kinds of fibers, fillers, and nanomaterials in this situation, the engineering potential of polymer-based composite materials was improved. In this context, research into the creation of laminated polymer nanocomposite (LPNC) materials is still in its early stages and calls for more involvement from academic, research, industrial, and institutional participants.

1.4.1. Fibers

Fiber is an essential and efficient material for the manufacture of improved composite materials. Fiber polymer composites (LPC) are being extensively used as lightweight structural materials in many components. Carbon fibers are about 5-10 micrometers in diameter and are mainly constituted of carbon atoms, as displayed in Figure 1.5 (Bhatt & Goe, 2017). The properties of carbon fiber include high-level stiffness, tensile strength, high-temperature tolerance, reduced weight, chemical stability, and minimal thermal expansion (Bhatt & Goe, 2017; Newcomb & Chae, 2018).

Figure 1.5. Carbon fiber.

1.4.2. Carbon Nanomaterials (CNMs)

Carbon nanomaterials (CNMs) are extensively used to reinforce polymer composites due to improved chemical and mechanical features. Under CNMs, there is a variety of potential nanomaterials categorized, such as CNT, CNO, CNR, GO, rGO, etc (Garriga et al., 2020; Patel et al., 2019; Zaytseva & Neumann, 2016). The GO consists of a high aspect ratio, surface area, and dispersion rate properties, which are highly required for synergistic effect (Aboutalebi et al., 2011). Its rapid growth in susceptible, efficient, low-power consumption, sensors, hence novel nanomaterials are being explored for multifunctional applications (Abdel-Karim et al., 2020; Alam et al., 2016; Ghorai, 2013; Rezk et al., 2020; Son & Bao, 2018). The development of hybrid materials using different combinations of reinforcement materials is one ultimate goal for a versatile number of studies in science, engineering, and the medical field. This chapter deals with the methodology of the development and characterization of polymer nanocomposite modified by GO/CF. After the fabrication phase, the developed nanocomposites are tested for mechanical characterization and other morphology investigations to evaluate the feasibility of the composite in depth. CNMs have high mechanical strength, thermal and electrical conductivity, and optical characteristics. The use of such materials for making a variety of products, including electrical equipment with increased strength, has been thoroughly studied. Figure 1.6 classifies the carbon nanomaterials (CNMs).:

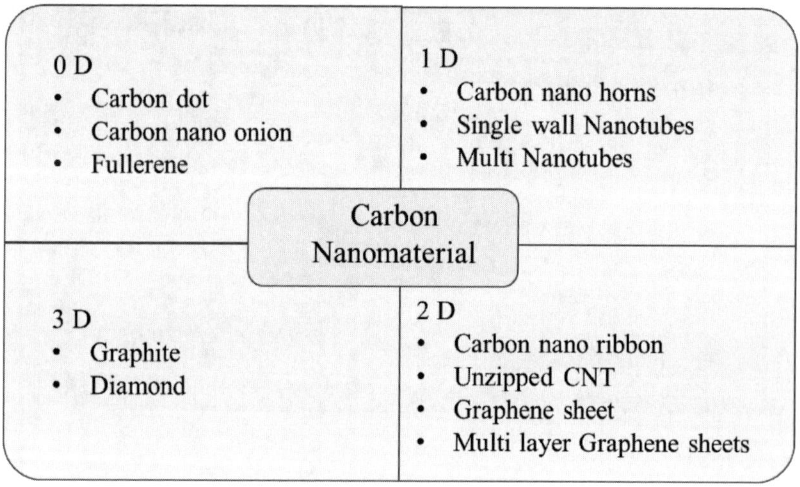

Figure 1.6. Categorization of carbon nanomaterials.

1.5. Properties of Carbon Nanomaterials (CNMs)

Various scientists have been finding innovative ways to explore the exceptional features of Graphene for a few decades. Phillip Wallace, a Canadian physician, proposed his innovative graphite electrical behaviour theory in 1947. Hensn-Peter Behmn and colleagues developed a single-layer graphite structure in a liquid combination with a low concentration in 1961 (Boehm et al., 1962). However, Geim and Novoselov discovered graphene (2004) (Novoselov et al., 2005) (Yanwu Zhu, Hengxing Ji, Hui-Ming Cheng, 2018). As a result, graphene is characterized as a two-dimensional (2D) structure of carbon particles. It is constructed similarly to a block of carbon material (Y. Li et al., 2012; Y. Zhu et al., 2010). Due to Graphene's excellent qualities, many such materials can be replaced in current use. Owing to several severe qualities being united in a single substance, graphene is also recognized as a component of cutting-edge technologies (Cheng et al., 2014). Owning properties such as transparency, conductivity, and elasticity will be helpful in adapting electronics to new environments (Nieto et al., 2017; Prashantha Kumar & Anthony Xavior, 2014). Figure 1.7 depicts the anticipated 2D carbon allotrope structure of graphene, which consists of atom vertices in a lattice with sp2 hybridization and a distance between carbon atoms of about 0.142 nm (De Andres et al., 2008). Figure 1.8 demonstrates that the conductivity of Graphene, which is remarked as efficient, thereby differentiates the methods.

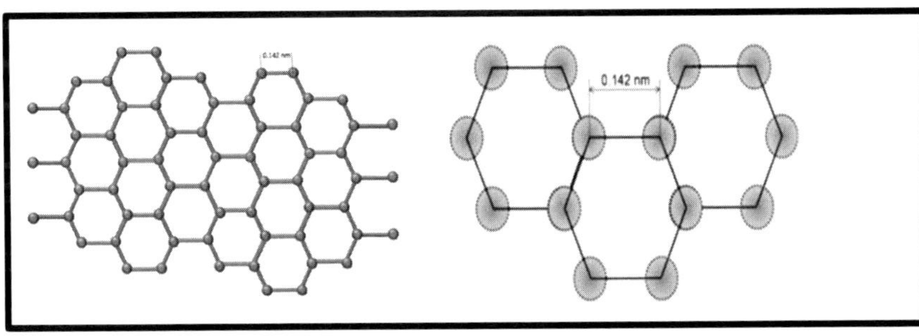

Figure 1.7. Graphene lattice structure of a honeycomb carbon atom in 0.142 nm length (Zhen & Zhu, 2017).

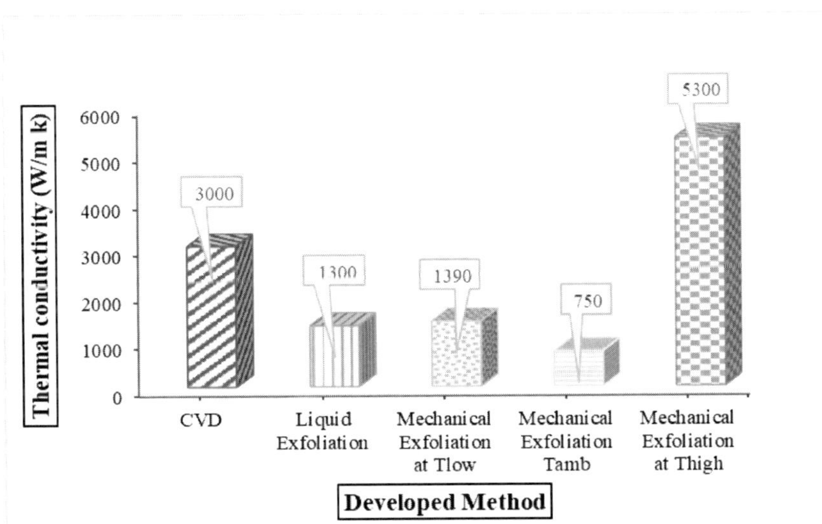

Figure 1.8. Overview of the thermal conductivity of graphene oxide (Cai et al., 2010; Chen et al., 2011; Novoselov et al., 2012) (Pei & Cheng, 2012) (Renteria et al., 2015) (P. Kumar et al., 2015) (J.-U. Lee, Yoon, Kim, Lee, & Cheong, 2011).

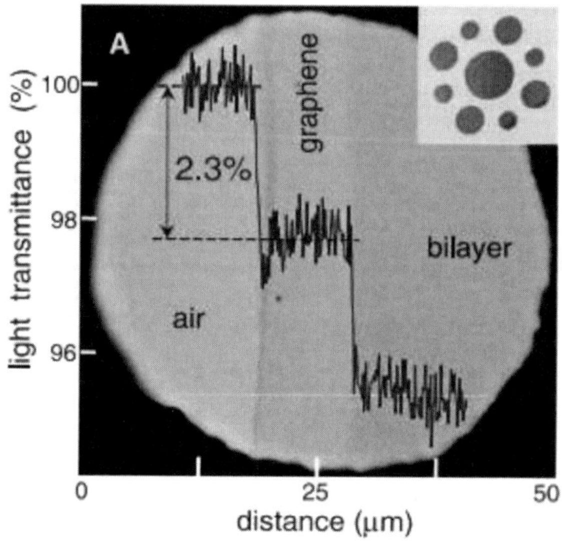

Figure 1.9. Transmittance of white light function (Zhen & Zhu, 2017).

The smooth structural constant = 0.0072, also known as the coefficient of absorption, was hypothesized to exist in nature by the optical properties of graphene, which respond like an electron, including its spin (Xu, 2017). Without a microscope, it is visible and can absorb 2.3 percent of white light. On a silicon substrate, optical microscopy may also identify graphene. The absorption spectrum shows a clear contrast between the various layers of graphene (Zhen & Zhu, 2017). The condensed properties of a physical substance are determined by a fundamental constant that has been established and is independent of the material parameter (Mohan et al., 2018). Figure 1.9 shows that only one-atom-thick graphene absorbs substantial incident light. Nanomaterial offers superior thermal characteristics and high tensile strength, fatigue, and creep resistivity compared to epoxy polymer (Sengupta et al., 2011). Relative to neat epoxy, the graphene material has superior tensile strength, fatigue, creep, and thermal properties.

1.6. Introduction to Laminated Polymer Nanocomposites

When it comes to mechanical strength, composite materials based on polymers often fail to meet in applications. Therefore, several synthetic and man-made filler/nanomaterials were used to increase the mechanical strength of polymer-based composite materials (Kostagiannakopoulou et al., 2017; Pathak et al., 2019; Shagor et al., 2021). The kind of fillers can improve the LPC product required manufacturing capabilities in the epoxy (matrix) (Hosur et al., 2018; Kunrath et al., 2019). To put nanoparticles and the unique features they offer to good use in practical, multipurpose contexts, polymer nanocomposites have been developed (M. Kumar et al., 2020; Siwal et al., 2020). Material for the fabrication of high-performance composites should have higher mechanical, electrical, and thermal properties. On the latter, the laminate has excellent interfacial bonding, is longer and more proximal, and has a high tensile strength (Shin et al., 2017). A longer, highly oriented laminate along the strain and nano filler axis is therefore necessary for greater mechanical performance, thermal conductivity, and electrical conductivity (Gangineni et al., 2019). The interface between the nanofiller epoxy bonding and the fibre (strength, width) determines the laminate's shear strength. Additionally, overall performance is impacted by the degree of randomness of the nano filler in the matrix and the laminate's bonding strength in nanomaterials epoxy (Naebe et al., 2014). Therefore, it requires very particular values for tensile, shear, and bending strength. Mechanical strength is increased in comparison to plain materials, and this is one of the main advantages of LPNC. As a result, the high potential properties of LPNC distinguish it from metals and their alloys. It is suitable for manufacturing multifunctional component development. Higher stiffness, thermal, and mechanical qualities can be achieved with the use of cross-linked connections in addition to thermoset polymers. Nanofiller materials reinforcing epoxy resin increased stiffness and strength under varied applied loads. The epoxy matrix's viable nanomaterial improves flexural behaviour (Burkov & Eremin, 2018; Gang, 2018). Therefore, nanomaterials enhance macro-reinforced (fiber) composite characteristics. The doping concept is used to study polymers' physical and mechanical properties in fiber composites. Nanomaterials in Carbon/epoxy composites have just recently been an active area of research. Polymer nanocomposites need additional study before they can be used effectively in multipurpose items.

Composite laminates are increasingly being employed in a variety of emerging technologies because of their distinctive mechanical and physical characteristics. These include the automotive sector, sports equipment, aerospace items, ship marine, and the petroleum business (Mohammed et al., 2015; Wang et al., 2011). Glass, carbon, and Kevlar laminate are the primary focus of research into LPC, with an emphasis on the utilization of products that have a variety of advantages (Nijssen & R. P. L. Nijssen, 2015). It is recommended that the primary structural elements of aircraft be replaced with produced LPC material to reduce weight by about 50 percent (A. Kumar et al., 2019; Myagkov et al., 2014). The utilization of composites that are both lightweight and high strength is essential for the advancement of engine performance and reduced fuel consumption in the design of aircraft for future growth (Nonobe, 2017; Stewart, 2009). Polymer composites are used for more purposes than metals because of their outstanding mechanical qualities, including low weight, stiffness to weight, corrosion, fatigue, and heat coefficient (Myagkov et al., 2014; Nguyen et al., 2019; Stewart, 2009). Laminates can be tailored to specific applications by changing carbon fibre ratios and the orientation of each ply (Liu et al., 2012). The reinforced component can be comprised of a variety of different materials and methods, such as particles, flakes, whiskers, laminates, or fibers (George et al., 2001; Mohammed et al., 2015).

1.7. Advantages and Applications of Laminated Polymer Nanocomposites

It is typically for the following reasons that composites are chosen over traditional materials such as metal alloys or wood (Hung, 2006). Figure 1.10 presents the benefits of composite materials over traditional materials.

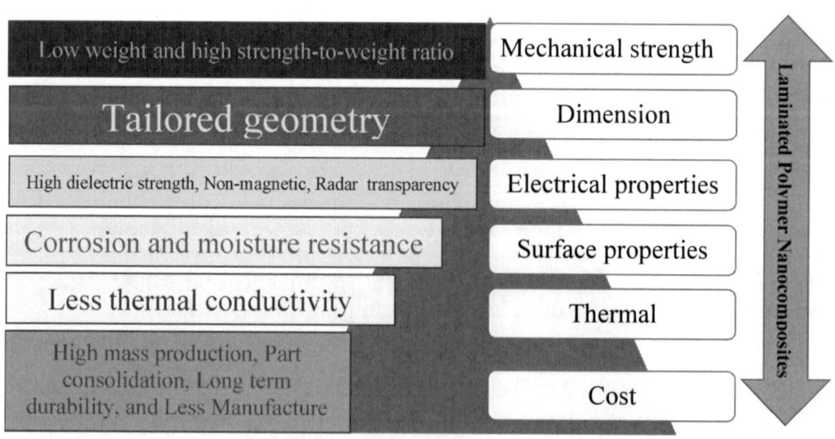

Figure 1.10. Advantage of composite material.

Also, it has various advantages:

- Higher specific strength than metals, non-metals, and even alloys
- Lower specific gravity
- Superior stiffness
- Improved toughness
- Fabrication or production is cheaper
- Fatigue strength and creep are superior.
- Controlled electrical conductivity
- Corrosion resistance.

Figure 1.11. Application of CFRP composite (Muhammad et al., 2021).

Transportation, automobiles, aerospace, defence, sports products, and infrastructure comprise the primary applications of polymer nanocomposite materials (Hanemann & Szabó, 2010; Hazra & Basu, 2016; Javidparvar et al., 2019; A. Kumar et al., 2019). As a result of its exceptional corrosion resistance, high strength-to-weight ratio, moisture resistance, ease of design, and flexibility in the handling process. The epoxy made from nanomaterial-modified carbon fiber-reinforced plastics (CFRP) demonstrates a variety of exceptional qualities, including reduced weight, high strength, rigidity, durability, impact tolerance, resistance to moisture, and chemical stabilisation. As a result, it is extensively employed in the production of efficient aerospace appliances (fuselages, limbs, ribs, panels, fuel tanks, and tail assemblies), as well as building products and automobiles (Müller et al., 2017). In the series, CFRP epoxy composites that have been modified with graphene oxide (GO) are employed in the conrod of the R4 (84/71) engine and the piston rod (Das & Prusty, 2013), high-pressure hydrogen tank (Siwal et al., 2020), cryo tank components (Baetcke & Kaltschmitt, 2018). Ford Motors has recently employed graphene to create engineered components that generate noise, including belt-driven pulleys, fuel rails, pumps, chain drivers, and belt covers (Ali et al., 2020; Nguyen Bich & Nguyen Van, 2016). Additionally, the manufacturing of spars, hulls, tills, rudders, keels, masts, and long upright and parallel (deck) poles of marine components is also facilitated by graphene oxide-modified CFRP (Jiang et al., 2019; Jin et al., 2021).

The application of polymer nanocomposites is summarized and displayed in Figure 1.11.

- Aerospace Industry, Aeronautics, Automobiles, and Aircraft wings industries (Long et al., 2019)
- Electronics Packaging, electronics application (Phiri et al., 2017)
- Electromagnetic interface shielding (A. Li et al., 2017)
- Fuel cells and their Bipolar Plates (Rzeczkowski et al., 2019)
- Wind turbine blades (Pathak et al., 2016)
- Sporting goods, rackets, and golf sticks (Choudhary & Gupta, 2001)
- Bio-medical (Choudhary & Gupta, 2001)
- Academic area (M. Lee et al., 2015)

1.8. Machining of Laminated Polymer Nanocomposites (LNPC)

In comparison to metallic materials and their mixtures, LPC cannot be machined effectively. For polymers and their composites, the machining phenomenon of metals is not applicable. The components of manufactured polymer composites are primarily produced in desired shapes. However, fitting and joining into the final assembly of various items still requires a secondary manufacturing procedure (Fleischer et al., 2018; Slamani et al., 2015). Drilling, milling, turning, and other primary machining operations are used on polymers to produce effective structures. However, research into the manufacture of polymers in the past shows that machining polymers is complex. This complexity is primarily caused by anisotropy, the non-homogeneous structure of the polymer, and the high abrasion resistance of the reinforcing material. This causes a wide variety of flaws and wear that may be traced back to the machining process. This could potentially impact the life of the tool and the cutting forces. Research indicates that the formation of defects and damage during machining can be substantially diminished by carefully controlling machining constraints (Ozkan et al., 2019; Voss et al., 2017). Polymers reinforced with carbon nanomaterials (CNMs) such as carbon nanorods/ nanotubes/ graphene/ reduced-graphene-oxide, etc., present significant hurdles during machining for both industry and researchers due to their complex structure and behavior. Both the manufacturing industry and academic institutions face a difficult challenge in optimizing tool life and balancing cutting conditions for product production (Roy et al., 2017; Sengupta et al., 2011). In the production of polymers, drilling is one of the most fundamental machining processes, the analysis of which can be informed by attention to optimal working conditions and tool design (Caggiano, 2018). The operating conditions of the tool materials and tool shape must be designed to minimize heat generation to achieve the desired machining performances, including high cutting rates, low tool attrition, and surface integrities. This makes drilling holes in polymer composites a challenging machining task. Conceptually, drilling is the process of using a machine to remove metal and create a neat hole (Figure 1.12). The plastic fabric has undergone an optimal machining technique, making it possible for the part to be mechanically attached and installed. The quality of the hole was influenced by the tool design and other process parameters when drilling polymeric materials. Researchers from a variety of fields have examined polymer composite materials and their machining processes. Many experts in the field evaluated the performance, accuracy, and delamination of drilling in LPC material. Epoxy composite material presents significant obstacles during machining for both academics and industry due to its complex structure and behaviour. Cutting force and tool life in situations with variable parameters also present challenges for manufacturers. The tool revolves in the y-axis of the workpiece while feeding in the x-axial direction according to the cutting theory of the milling operation (Figure 1.13).

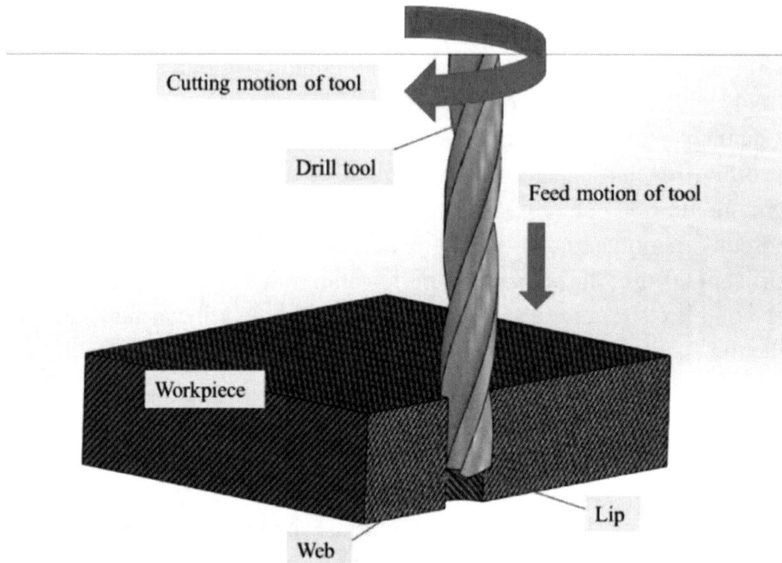

Figure 1.12. Principle of the drilling process.

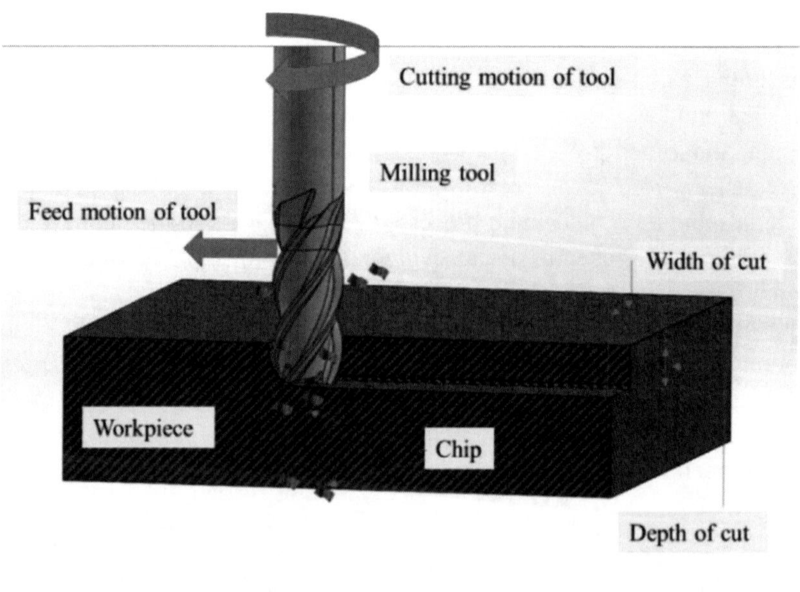

Figure 1.13. Principle of the milling process.

Conclusion

As a result of their exceptional mechanical, thermal, and electrical properties, this chapter emphasizes composite materials, their classification, and similar critical functions in many engineering disciplines. It demonstrates that laminated polymer composites (LPC) can be used in a variety of engineering applications. The fundamentals of machining studies applied to laminated polymer nanocomposites are discussed. After the brief introduction of laminated polymer nanocomposites (LNPC), their application and necessity in the field of engineering will be addressed. The next chapter consists of fabrication processes and characterization analysis (Mechanical, Physical, Thermal, Electrical, and Morphology) of LNPC.

References

Abdel-Karim, R., Reda, Y., & Abdel-Fattah, A. (2020). Review—Nanostructured Materials-Based Nanosensors. *Journal of The Electrochemical Society*, *167*(3), 037554.

Aboutalebi, S. H., Gudarzi, M. M., Zheng, Q. Bin, & Kim, J. K. (2011). Spontaneous formation of liquid crystals in ultralarge graphene oxide dispersions. *Advanced Functional Materials*, *21*(15), 2978–2988.

Alam, A., Meng, Q., Shi, G., Arabi, S., Ma, J., Zhao, N., & Kuan, H. C. (2016). Electrically conductive, mechanically robust, pH-sensitive graphene/polymer composite hydrogels. *Composites Science and Technology*, *127*, 119–126.

Ali, M. K. A., Hou, X., & Abdelkareem, M. A. A. (2020). Anti-wear properties evaluation of frictional sliding interfaces in automobile engines lubricated by copper/graphene nanolubricants. *Friction*, *8*(5), 905–916. https://doi.org/10.1007/s40544-019-0308-0.

Ashby, M. F. (2013). Material selection strategies. In *Materials and the Environment* (2nd ed., pp. 227–273). Elsevier Inc.

Ashby, M. F., Bréchet, Y. J. M., Cebon, D., & Salvo, L. (2004). Selection strategies for materials and processes. *Materials and Design*, *25*(1), 51–67.

Baetcke, L., & Kaltschmitt, M. (2018). Hydrogen storage for mobile application: Technologies and their assessment. In C. Azzaro-Pantel (Ed.), *Hydrogen Supply Chains Design, Deployment and Operation* (pp. 167–206). Elsevier Ltd. https://doi.org/10.1016/B978-0-12-811197-0.00005-1.

Ballout, W., Sallem-Idrissi, N., Sclavons, M., Doneux, C., Bailly, C., Pardoen, T., & Van Velthem, P. (2022). High performance recycled CFRP composites based on reused carbon fabrics through sustainable mild solvolysis route. *Scientific Reports*, *12*(1), 5928. https://doi.org/10.1038/s41598-022-09932-0.

Basu, S., & Debnath, A. K. (2015). Boiler Control System. In *Power Plant Instrumentation and Control Handbook* (2nd ed., pp. 585–694). Elsevier Ltd. https://doi.org/10.1016/b978-0-12-800940-6.00008-3.

Bhatt, P., & Goe, A. (2017). Carbon Fibres: Production, Properties and Potential Use. *Material Science Research India, 14*(1), 52–57.

Boehm, H. P., Clauss, A., Fischer, G. O., & Hofmann, U. (1962). The Adsorption Behavior of Very Thin Carbon Films. *Zcitsclirift Fiir Anorganische Und Allgenieine Cheniie (ZAAC), 316*(3–4), 119–127.

Brebu, M. (2020). Environmental degradation of plastic composites with natural fillers-a review. *Polymers, 12*(1), 166 (1-22). https://doi.org/10.3390/polym12010166.

Burkov, M., & Eremin, A. (2018). Hybrid CFRP/SWCNT Composites with Enhanced Electrical Conductivity and Mechanical Properties. *Journal of Materials Engineering and Performance, 27*, 5984–5991.

Caggiano, A. (2018). Machining of fibre reinforced plastic composite materials. *Materials, 11*(3), 442 (1-20). https://doi.org/10.3390/ma11030442.

Cai, W., Moore, A. L., Zhu, Y., Li, X., Chen, S., Shi, L., & Ruoff, R. S. (2010). Thermal transport in suspended and supported monolayer graphene grown by chemical vapor deposition. *Nano Letters, 10*(5), 1645–1651.

Cao, K., Feng, S., Han, Y., Gao, L., Ly, T. H., Xu, Z., & Lu, Y. (2020). Elastic straining of free-standing monolayer graphene. *Nature Communications, 11*(1), 1–7.

Chen, S., Moore, A. L., Cai, W., Suk, J. W., An, J., Mishra, C., Amos, C., Magnuson, C. W., Kang, J., Shi, L., & Ruoff, R. S. (2011). Raman measurements of thermal transport in suspended monolayer graphene of variable sizes in vacuum and gaseous environments. *ACS Nano, 5*(1), 321–328.

Cheng, H., Hu, C., Zhao, Y., & Qu, L. (2014). Graphene fiber: A new material platform for unique applications. *NPG Asia Materials, 6*(7), e113 (1-13).

Choudhary, V., & Gupta, A. (2001). Polymer/Carbon Nanotube Nanocomposites. In S. Yellampalli (Ed.), *Carbon Nanotubes - Polymer Nanocomposites* (Issue July 2015, pp. 65–90). Intech Open. https://doi.org/10.5772/18423.

Cui, M., Ren, S., Pu, J., Wang, Y., Zhao, H., & Wang, L. (2019). Poly (o-phenylenediamine) modified graphene toward the reinforcement in corrosion protection of epoxy coatings. *Corrosion Science, 159*, 108131.

Das, T. K., & Prusty, S. (2013). Graphene-Based Polymer Composites and Their Applications. *Polymer - Plastics Technology and Engineering, 52*(4), 319–331.

De Andres, P. L., Ramírez, R., & Vergés, J. A. (2008). Strong covalent bonding between two graphene layers. *Physical Review B - Condensed Matter and Materials Physics, 77*(4), 1–5.

Findik, F., & Turan, K. (2012). Materials selection for lighter wagon design with a weighted property index method. *Materials and Design, 37*, 470–477.

Fleischer, J., Teti, R., Lanza, G., Mativenga, P., Möhring, H. C., & Caggiano, A. (2018). Composite materials parts manufacturing. *CIRP Annals, 67*(2), 603–626.

Gang, D. (2018). The effect of surface treatment of CF and graphene oxide on the mechanical properties of PI composite. *Journal of Thermoplastic Composite Materials, 31*(9), 1219–1231. https://doi.org/10.1177/0892705717734606.

Gangineni, P. K., Yandrapu, S., Ghosh, S. K., Anand, A., Prusty, R. K., & Ray, B. C. (2019). Mechanical behavior of Graphene decorated carbon fiber reinforced polymer composites: An assessment of the influence of functional groups. *Composites Part A: Applied Science and Manufacturing*, *122*, 36–44. https://doi.org/10.1016/j.compositesa.2019.04.017.

Garriga, R., Herrero-Continente, T., Palos, M., Cebolla, V. L., Osada, J., Muñoz, E., & Rodríguez-Yoldi, M. J. (2020). Toxicity of carbon nanomaterials and their potential application as drug delivery systems: In vitro studies in caco-2 and mcf-7 cell lines. *Nanomaterials*, *10*(8), 1–21.

Gay, D., Hoa, S. V, & Tsai, S. W. (2002). Composite Materials Design and Applications. In D. Gay, S. V. Hoa, & S. W. Tsai (Eds.), *Taylor and Francis* (3rd ed.). CRC Press. https://doi.org/10.1201/9781420031683.

George, J., Sreekala, M. S., & Thomas, S. (2001). A review on interface modification and characterization of natural fiber reinforced plastic composites. *Polymer Engineering and Science*, *41*(9), 1471–1485. https://doi.org/10.1002/pen.10846.

Ghorai, S. (2013). Chemical, physical and mechanical properties of nanomaterials and its applications [University of Iowa]. In *University of Iowa*. https://doi.org/10.17077/etd.5r1xf0y8.

Hanemann, T., & Szabó, D. V. (2010). Polymer-nanoparticle composites: From synthesis to modern applications. *Materials*, *3*(6), 3468–3517. https://doi.org/10.3390/ma3063468.

Hazra, S., & Basu, S. (2016). Graphene-Oxide Nano Composites for Chemical Sensor Applications. *C — Journal of Carbon Research*, *2*(2), 12. https://doi.org/10.3390/c2020012.

Horoschenkoff, A., & Christner, C. (2012). Carbon Fibre Sensor: Theory and Application. In N. Hu (Ed.), *Composites and Their Applications* (p. 3484). Intech Open. https://doi.org/10.5772/50504.

Hosur, M., Mahdi, T., & Jeelani, S. (2018). Studies on the performance of multi-phased carbon/epoxy composites with nanoclay and multi-walled carbon nanotubes. *Multiscale and Multidisciplinary Modeling, Experiments and Design*, *1*(4), 255–268. https://doi.org/10.1007/s41939-018-0017-9.

Hung, C.-C. (2006). *Carbon Materials Metal/Metal Oxide Nanoparticle Composite and Battery Anode Composed of the Same, US-Patent-7,094,499 B1.*

Jannatyha, N., Shojaee-Aliabadi, S., Moslehishad, M., & Moradi, E. (2020). Comparing mechanical, barrier and antimicrobial properties of nanocellulose/CMC and nanochitosan/CMC composite films. *International Journal of Biological Macromolecules*, *164*, 2323–2328. https://doi.org/10.1016/j.ijbiomac.2020.07.249.

Javidparvar, A. A., Naderi, R., & Ramezanzadeh, B. (2019). Epoxy-polyamide nanocomposite coating with graphene oxide as cerium nanocontainer generating effective dual active/barrier corrosion protection. *Composites Part B: Engineering*, *172*, 363–375. https://doi.org/10.1016/j.compositesb.2019.05.055 Received.

Jiang, T., Qi, L., & Qin, W. (2019). Improving the Environmental Compatibility of Marine Sensors by Surface Functionalization with Graphene Oxide [Research-article]. *Analytical Chemistry*, *91*(20), 13268–13274. https://doi.org/10.1021/acs.analchem.9b03974.

Jin, H., Tian, L., Bing, W., Zhao, J., & Ren, L. (2021). Toward the Application of Graphene for Combating Marine Biofouling. *Advanced Sustainable Systems*, *5*(1), 1–16. https://doi.org/10.1002/adsu.202000076.

Kato, A., Ikeda, Y., & Kohjiya, S. (2012). Carbon Black-Filled Natural Rubber Composites: Physical Chemistry and Reinforcing Mechanism. In S. Thomas, K. Joseph, S. K. Malhotra, K. Goda, & M. S. Sreekala (Eds.), *Polymer Composites* (1st ed., Vol. 1, pp. 515–543). Wiley.

Koch, D., Tushtev, K., Horvath, J., Knoche, R., & Grathwohl, G. (2006). Evaluation of Mechanical Properties and Comprehensive Modeling of CMC with Stiff and Weak Matrices. *Advances in Science and Technology*, *45*, 1435–1443. https://doi.org/10.4028/www.scientific.net/AST.45.1435.

Kostagiannakopoulou, C., Tsilimigkra, X., Sotiriadis, G., & Kostopoulos, V. (2017). Synergy effect of carbon nano-fillers on the fracture toughness of structural composites. *Composites Part B: Engineering*, *129*, 18–25.

Kumar, A., Sharma, K., & Dixit, A. R. (2019). A review of the mechanical and thermal properties of graphene and its hybrid polymer nanocomposites for structural applications. *Journal of Materials Science*, *54*(8), 5992–6026. https://doi.org/10.1007/s10853-018-03244-3.

Kumar, M., Saini, J. S., & Bhunia, H. (2020). Performance of Mechanical Joints Prepared from Carbon-Fiber-Reinforced Polymer Nanocomposites under Accelerated Environmental Aging. *Journal of Materials Engineering and Performance*, *29*, 7511–7525. https://doi.org/10.1007/s11665-020-05216-8.

Kunrath, K., Kerche, E. F., Rezende, M. C., & Amico, S. C. (2019). Mechanical, electrical, and electromagnetic properties of hybrid graphene/glass fiber/epoxy composite. *Polymers and Polymer Composites*, *27*(5), 1–6. https://doi.org/10.1177%2F0967391119828559.

Lee, C., Wei, X., Kysar, J. W., & Hone, J. (2008). Measurement of the elastic properties and intrinsic strength of monolayer graphene. *Science*, *321*(5887), 385–388.

Lee, M., Lee, J., Park, S. Y., Min, B., Kim, B., & In, I. (2015). Production of graphene oxide from pitch-based carbon fiber. *Scientific Reports*, *5*, 1–10. https://doi.org/10.1038/srep11707.

Li, A., Zhang, C., & Zhang, Y. F. (2017). Thermal conductivity of graphene-polymer composites: Mechanisms, properties, and applications. *Polymers*, *9*(9), 437 (1-17). https://doi.org/10.3390/polym9090437.

Li, Y., Chem, J. M., Peng, Q., He, X., Hu, P., Wang, C., Shang, Y., Wang, R., Jiao, W., Lv, H., & Chem, J. M. (2012). Synthesis and characterization of a new hierarchical reinforcement by chemically grafting graphene oxide onto carbon fibers. *Journal of Materials Chemistry*, *22*(36), 18748–18752.

Liu, D. F., Tang, Y. J., & Cong, W. L. (2012). A review of mechanical drilling for composite laminates. *Composite Structures*, *94*(4), 1265–1279. https://doi.org/10.1016/j.compstruct.2011.11.024.

Long, J. P., Li, S. X., Liang, B., & Wang, Z. G. (2019). Investigation of thermal behaviour and mechanical property of the functionalised graphene oxide/epoxy resin nanocomposites. *Plastics, Rubber and Composites*, *48*(3), 127–136. https://doi.org/10.1080/14658011.2019.1577027.

Mohammed, L., Ansari, M. N. M., Pua, G., Jawaid, M., & Islam, M. S. (2015). A Review on Natural Fiber Reinforced Polymer Composite and Its Applications. *International Journal of Polymer Science*, *2015*, 243947 (1-15). https://doi.org/10.1155/2015/243947.

Mohan, V. B., Lau, K. tak, Hui, D., & Bhattacharyya, D. (2018). Graphene-based materials and their composites: A review on production, applications and product limitations. *Composites Part B: Engineering*, *142*, 200–220.

Muhammad, A., Rahman, Md. R., Baini, R., & Bin Bakri, M. K. (2021). Applications of sustainable polymer composites in automobile and aerospace industry. In Md. R. Rahman (Ed.), *Advances in Sustainable Polymer Composites* (pp. 185–207). Elsevier Ltd. https://doi.org/10.1016/b978-0-12-820338-5.00008-4.

Müller, K., Bugnicourt, E., Latorre, M., Jorda, M., Sanz, Y. E., Lagaron, J. M., Miesbauer, O., Bianchin, A., Hankin, S., Bölz, U., Pérez, G., Jesdinszki, M., Lindner, M., Scheuerer, Z., Castelló, S., & Schmid, M. (2017). Review on the processing and properties of polymer nanocomposites and nanocoatings and their applications in the packaging, automotive and solar energy fields. *Nanomaterials*, *7*(4), 74.

Myagkov, L. L., Mahkamov, K., Chainov, N. D., & Makhkamova, I. (2014). Advanced and conventional internal combustion engine materials. In R. Folkson (Ed.), *Alternative Fuels and Advanced Vehicle Technologies for Improved Environmental Performance: Towards Zero Carbon Transportation* (pp. 370–392). Woodhead Publishing Limited. https://doi.org/10.1533/9780857097422.2.370.

Naebe, M., Wang, J., Amini, A., Khayyam, H., Hameed, N., Li, L. H., Chen, Y., & Fox, B. (2014). Mechanical Property and Structure of Covalent Functionalised Graphene/Epoxy Nanocomposites. *Scientific Reports*, *4*, 4375 (1-7). https://doi.org/10.1038/srep04375.

Newcomb, B. A., & Chae, H. G. (2018). The properties of carbon fibers. In A. R. Bunsell (Ed.), *Handbook of Properties of Textile and Technical Fibres* (pp. 841–871). Woodhead Publishing Limited. https://doi.org/10.1016/B978-0-08-101272-7.00021-3.

Nguyen Bich, H., & Nguyen Van, H. (2016). Promising applications of graphene and graphene-based nanostructures. *Advances in Natural Sciences: Nanoscience and Nanotechnology*, *7*(2), 023002. https://doi.org/10.1088/2043-6262/7/2/023002.

Nguyen, T. A., Nguyen, Q. T., & Bach, T. P. (2019). Mechanical Properties and Flame Retardancy of Epoxy Resin/Nanoclay/Multiwalled Carbon Nanotube Nanocomposites. *Journal of Chemistry*, *2019*, 3105205 (1-9). https://doi.org/10.1155/2019/3105205.

Nieto, A., Bisht, A., Lahiri, D., Zhang, C., & Agarwal, A. (2017). Graphene reinforced metal and ceramic matrix composites: a review. *International Materials Reviews*, *62*(5), 241–302.

Nijssen, R. P. L., & R. P. L. Nijssen. (2015). Composite Materials an introduction. In R. P. L. Nijssen (Ed.), *A VKCN publication* (2015th ed., Vol. 1). Inholland University of Applied Sciences.

Nonobe, Y. (2017). Development of the fuel cell vehicle mirai. *IEEJ Transactions on Electrical and Electronic Engineering*, *12*(1), 5–9. https://doi.org/10.1002/tee.22328.

Novoselov, K. S., Fal'Ko, V. I., Colombo, L., Gellert, P. R., Schwab, M. G., & Kim, K. (2012). A roadmap for graphene. *Nature*, *490*(7419), 192–200.

Novoselov, K. S., Geim, A. K., Morozov, S. V., Jiang, D., Katsnelson, M. I., Grigorieva, I. V., Dubonos, S. V., & Firsov, A. A. (2005). Two-dimensional gas of massless Dirac fermions in graphene. *Nature*, *438*, 197–200.

Ozkan, D., Sabri Gok, M., Oge, M., & Cahit Karaoglanli, A. (2019). Milling Behavior Analysis of Carbon Fiber-Reinforced Polymer (CFRP) Composites. *Materials Today: Proceedings*, *11*, 526–533.

Patel, K. D., Singh, R. K., & Kim, H. W. (2019). Carbon-based nanomaterials as an emerging platform for theranostics. *Materials Horizons*, *6*(3), 434–469.

Pathak, A. K., Borah, M., Gupta, A., Yokozeki, T., & Dhakate, S. R. (2016). Improved mechanical properties of carbon fiber/graphene oxide-epoxy hybrid composites. *Composites Science and Technology*, *135*, 28–38. https://doi.org/10.1016/j.compscitech.2016.09.007.

Pathak, A. K., Garg, H., Singh, M., Yokozeki, T., & Dhakate, S. R. (2019). Enhanced interfacial properties of graphene oxide incorporated carbon fiber reinforced epoxy nanocomposite: a systematic thermal properties investigation. *Journal of Polymer Research*, *26*(2), 23 (1-23). https://doi.org/10.1007/s10965-018-1668-2.

Phiri, J., Gane, P., & Maloney, T. C. (2017). General overview of graphene: Production, properties and application in polymer composites. *Materials Science and Engineering B*, *215*, 9–28. https://doi.org/10.1016/j.mseb.2016.10.004.

Prashantha Kumar, H. G., & Anthony Xavior, M. (2014). Graphene reinforced metal matrix composite (GRMMC): A review. *Procedia Engineering*, *97*, 1033–1040.

Ray, S., & Easteal, A. J. (2007). Advances in polymer-filler composites: Macro to nano. *Materials and Manufacturing Processes*, *22*(6), 741–749.

Rezk, M. Y., Sharma, J., & Gartia, M. R. (2020). Engineered Nanomaterial for Industrial Use. *Nanomaterials Review*, *10*(2251), 10112251.

Roy, Y. A., Gobivel, K., Sekar, K. S. V., & Kumar, S. S. (2017). Impact of cutting forces and chip microstructure in high speed machining of carbon fiber - Epoxy composite tube. *Archives of Metallurgy and Materials*, *62*(3), 1771–1777.

Rubino, F., Nisticò, A., Tucci, F., & Carlone, P. (2020). Marine application of fiber reinforced composites: A review. *Journal of Marine Science and Engineering*, *8*(1), 26 (1-28). https://doi.org/10.3390/jmse8010026.

Rzeczkowski, P., Krause, B., & Pötschke, P. (2019). Characterization of highly filled PP/graphite composites for adhesive joining in fuel cell applications. *Polymers*, *11*(3), 462. https://doi.org/10.3390/polym11030462.

Sengupta, R., Bhattacharya, M., Bandyopadhyay, S., & Bhowmick, A. K. (2011). A review on the mechanical and electrical properties of graphite and modified graphite reinforced polymer composites. *Progress in Polymer Science (Oxford)*, *36*(5), 638–670. https://doi.org/10.1016/j.progpolymsci.2010.11.003.

Seydibeyoğlu, M. Ö., Doğru, A., Kandemir, M. B., & Aksoy, Ö. (2020). Lightweight Composite Materials in Transport Structures. *Lightweight Polymer Composite Structures*, *August*, 103–130. https://doi.org/10.1201/9780429244087-5.

Shagor, R. M. R., Abedin, F., & Asmatulu, R. (2021). Mechanical and thermal properties of carbon fiber reinforced composite with silanized graphene as nano-inclusions.

Journal of Composite Materials, *55*(5), 597–608. https://doi.org/10.1177/0021998320953183.

Shin, P. S., Kwon, D. J., Kim, J. H., Lee, S. Il, DeVries, K. L., & Park, J. M. (2017). Interfacial properties and water resistance of epoxy and CNT-epoxy adhesives on GFRP composites. *Composites Science and Technology*, *142*, 98–106. https://doi.org/10.1016/j.compscitech.2017.01.026.

Singh, K. (2014). Advanced Materials for Land Based Gas Turbines. *Transactions of the Indian Institute of Metals*, *67*(5), 601–615. https://doi.org/10.1007/s12666-014-0398-3.

Siwal, S. S., Zhang, Q., Devi, N., & Thakur, V. K. (2020). Carbon-based polymer nanocomposite for high-performance energy storage applications. *Polymers*, *12*(3), 505 (1-31). https://doi.org/10.3390/polym12030505.

Slamani, M., Gauthier, S., & Chatelain, J. F. (2015). A study of the combined effects of machining parameters on cutting force components during high speed robotic trimming of CFRPs. *Measurement: Journal of the International Measurement Confederation*, *59*, 268–283.

Son, D., & Bao, Z. (2018). Nanomaterials in Skin-Inspired Electronics: Toward Soft and Robust Skin-like Electronic Nanosystems. *ACS Nano*, *12*(12), 11731–11739.

Stewart, R. (2009). Lightweighting the automotive market. *Reinforced Plastics*, *53*(3), 14–21. https://doi.org/10.1016/S0034-3617(09)70078-5.

Sun, F., & Zhao, Y. P. (2020). Geomaterials evaluation: A new application of Ashby plots. *Materials (Basel)*, *13*(11), 2517.

Uygunoglu, T., Gunes, I., & Brostow, W. (2015). Physical and mechanical properties of polymer composites with high content of wastes including boron. *Materials Research*, *18*(6), 1188–1196. https://doi.org/10.1590/1516-1439.009815.

Voss, R., Seeholzer, L., Kuster, F., & Wegener, K. (2017). Influence of fibre orientation, tool geometry and process parameters on surface quality in milling of CFRP. *CIRP Journal of Manufacturing Science and Technology*, *18*, 75–91.

Wang, R.-M., Zheng, S.-R., & Zheng, Y.-P. (2011). Introduction to polymer matrix composites. In R.-M. Wang, S.-R. Zheng, & Y.-P. Zheng (Eds.), *Polymer Matrix Composites and Technology* (pp. 1–25). Woodhead Publishing Limited. https://doi.org/10.1533/9780857092229.1.

Xu, Z. (2017). Fundamental properties of graphene. In H. Zhu, Z. Xu, D. Xie, & Y. Fang (Eds.), *Graphene: Fabrication, Characterizations, Properties and Applications*. Elsevier Inc. https://doi.org/10.1016/B978-0-12-812651-6.00004-5

Yanwu Zhu, Hengxing Ji, Hui-Ming Cheng, and R. S. R. d. (2018). Mass Production and Industrial Applications of Graphene Materials. *National Science Review*, *5*(1), 90–101.

Zaytseva, O., & Neumann, G. (2016). Carbon nanomaterials: Production, impact on plant development, agricultural and environmental applications. *Chemical and Biological Technologies in Agriculture*, *3*(1), 1–26.

Zhen, Z., & Zhu, H. (2017). Structure and properties of graphene. In H. Zhu, Z. Xu, D. Xie, & Y. Fang (Eds.), *Graphene: Fabrication, Characterizations, Properties and Applications* (pp. 1–12). Elsevier Inc. https://doi.org/10.1016/B978-0-12-812651-6.00001-X.

Zhu, Y., Murali, S., Cai, W., Li, X., Suk, J. W., Potts, J. R., & Ruoff, R. S. (2010). Graphene and graphene oxide: Synthesis, properties, and applications. *Advanced Materials*, *22*(35), 3906–3924.

Chapter 2

Fabrication Processes and Characteristics of Hybrid Polymer Nanocomposites

2.1. Fabrication Process

The most common uses for polymer composites are in applications requiring a strong yet lightweight structure. In comparison to other materials, it is distinguished by its high strength-to-weight ratio, high corrosion resistance, adaptability in the manufacturing process, and simple design. As a result of its low resistance to Electrostatic discharge (ESD) and electromagnetic interference (EMI), it is commonly used as a protective material (Budka et al., 2010; Pande et al., 2014). In a fast-growing industry, though, this may not be enough. New, cheaper materials with better properties are needed. As a result, the qualities of polymer composites (structural materials) can be changed by adding nanomaterials (NMs) (J. Kumar, Kesarwani, et al., 2023; J. Kumar, Kumar, Jaiswal, et al., 2022). A uniform distribution (free of the agglomeration problem) is preferred for improved characteristics (Ding & Pacek, 2008; Hanemann & Szabó, 2010). Despite this, nanoparticles are particularly challenging to work with because their surface area makes it more likely for them to aggregate. During the manufacturing process, severe techniques are planned to keep carbon nanomaterials (CNMs) from agglomeration (Ashori et al., 2015a).

The inherent characteristics of nanoparticles are degraded, whereas surface functioning greatly enhances CNMs dispersion and stress transmission. Fragmentation and ineffective exfoliation are adverse effects of non-covalent performance over the covalent case (Gromov et al., 2005). The difficulty of the dispersal and stabilization problem is, therefore, readily apparent, and an appropriate choice of promises may be made considering the relevant applications. The superior properties of laminated polymer nanocomposites (LPNC) indicate that they are sustainable materials. For making layered LPNC, it is essential to add CNMs to the resin. Adhesion connections need to be improved so that significant amounts of load can be transferred between the CNMs and the matrix interface. NMs are infused into modified polymers and alternating layers of fibers to form LPNC. These materials exhibit improved engineering properties by combining the advantageous characteristics of nanoparticles and polymers. Here are various approaches for manufacturing LPNC that are commonly used.

2.1.1. Layer-by-Layer (LbL) Assembly

This technique involves the sequential deposition of alternating layers of polyelectrolytes and nanoparticles onto a substrate (De et al., 2020; Srivastava & Kotov, 2008; X. Zhang et al., 2007). The polyelectrolytes and nanoparticles are typically charged, allowing them to adsorb onto the substrate via electrostatic interactions. The layering process can be controlled by adjusting parameters such as pH, ionic strength, and deposition time.

2.1.2. Solvent Casting

In this method, a polymer solution is prepared by dissolving a polymer in a suitable solvent. Nanoparticles are dispersed within the polymer solution, and the mixture is cast onto a substrate. A solid film is subsequently left as the solvent evaporates. Multiple layers of polymer and nanocomposite films can be stacked to create a laminated structure (Armentano et al., 2011; Lin et al., 2011).

2.1.3. *In-Situ* Polymerization

This method involves the synthesis of the polymer in the presence of nanoparticles. The monomers and nanoparticles are mixed, and polymerization is initiated. The reaction can be carried out in bulk, solution, or suspension, depending on the specific system. The resulting polymer matrix incorporates the nanoparticles, forming a nanocomposite film that can be stacked to create a laminated structure (Althues et al., 2007; B. Chen & Evans, 2004; Grothe et al., 2012; Landry et al., 1992).

2.1.4. Electrostatic Layer-by-Layer (ELbL) Assembly

This technique is like LbL assembly but involves the use of electrostatic interactions between charged polymers and nanoparticles. The layers are deposited by applying an electric field, which drives the charged species toward the oppositely charged substrate. The process can be repeated to create a multilayered structure. These methods offer different advantages and can be chosen based on factors such as the desired properties, scalability, and compatibility with the polymer and nanoparticle materials being used. The selection of the appropriate fabrication method depends on the specific requirements of the LPNC (Artyukhin et al., 2004; Z. Li et al., 2011; Mcshane & Lvov, 2014).

2.1.5. Hand Layup

Hand layup, also known as a manual process, is used in composite manufacturing to fiber-reinforced composite parts. It is commonly employed for smaller-scale or custom production and is one of the oldest and most fundamental methods of fabrication of composite structures. In the hand layup process, prepare the mold. The mold is typically made of a rigid material such as fiberglass or carbon fiber, and it is often coated with a mold release agent to facilitate easy removal of the finished part. The layers of reinforcing materials, such as fiberglass or carbon fiber fabrics, are manually placed into the mold. These materials come in various forms, including woven fabric, chopped strands, or unidirectional tapes, depending on the specific requirements of the part being produced. The fibers are typically pre-cut or measured to fit the desired shape and size. Once the reinforcing materials are positioned in the mold, a liquid resin system is applied by hand. The resin can be polyester, epoxy, vinyl ester, or other suitable composite resins, depending on the desired properties of the final part. The resin is often mixed with a catalyst or hardener to initiate the curing process. Scholars employ brushes, rollers, or squeegees to ensure thorough wetting and impregnating the fibers. Additional materials, such as breather and vacuum bagging, may also be added to facilitate the consolidation of the laminate during the curing process. Depending on the resin system, the curing process can take place at room temperature or require the application of heat. The curing time varies depending on the resin and the size and complexity of the part (J. Kumar et al., 2021; J. Kumar, Kumar Verma, et al., 2020; J. Kumar, Verma, et al., 2020).

Hand layup gives more control over the shape of the part and can be used to make a lot of different composite parts, such as flat panels, curved surfaces, and parts with more complicated forms. However, it is a labor-intensive process that requires skilled technicians to ensure proper resin impregnation, fiber alignment, and overall quality. While hand layup is considered a manual and time-consuming process, it is still widely used in industries such as aerospace, automotive, marine, and sporting goods, particularly for low-volume or custom production where cost-effective methods are required or to produce large-scale composite structures. The hand layup process is demonstrated in Figure 2.1.

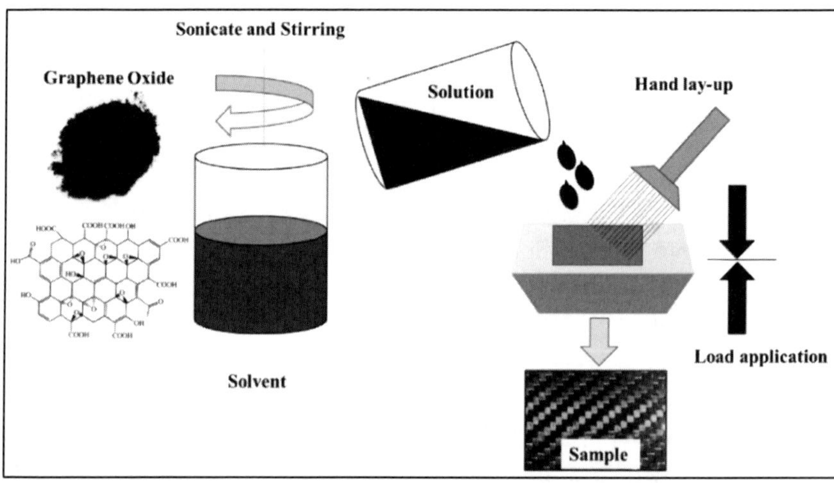

Figure 2.1. Hand layup preparation (J. Kumar, Verma, et al., 2020).

2.1.6. Vacuum-Assisted Resin Transfer Moulding (VARTM)

VARTM, also known as the vacuum infusion process, is used to fabricate high-performance composite structures (J. Kumar, Jaiswal, et al., 2023; J. Kumar, Kumar, Jaiswal, et al., 2022; J. Kumar, Kumar, Kumar, et al., 2022a). It is a variation of the resin transfer molding (RTM) process and is commonly employed in industries such as aerospace, automotive, marine, and wind energy. In VARTM, the process begins by placing a dry preform, consisting of fibers or fabric reinforcements, into a mold cavity. The mold is then closed, and a vacuum bag, along with a distribution medium, is carefully placed over the pre-form and sealed to create an airtight enclosure. Once the vacuum bag is in place, a vacuum is applied to the mold cavity, causing the bag to press against the pre-form tightly. The purpose of the vacuum is to remove air and ensure a controlled environment for the resin infusion process. Next, a liquid resin, typically a two-part epoxy or polyester resin, is introduced into the mold cavity. The resin is drawn into the pre-form under the influence of the vacuum pressure, gradually impregnating the fibers and displacing the air. The distribution medium helps evenly distribute the resin throughout the pre-form. As the resin fills the mold cavity, it progressively wets out the fibers and saturates the pre-form. The vacuum pressure helps to ensure that voids and air pockets are minimized, resulting in a more uniform and high-quality composite structure. After the resin infusion is complete, the part is left to cure under controlled temperature and pressure conditions. The curing process allows the resin to harden and bond the fibers together, creating a strong and durable composite component. VARTM offers several advantages over traditional open-mold processes. It enables precise control over resin flow, reduces the risk of voids and defects, and allows the production of complex shapes with consistent fiber-to-resin ratios. Additionally, VARTM can be a cost-effective method as it eliminates the need for autoclaves or high-pressure equipment typically required in other manufacturing processes. CNMs aggregate in the solvent blending due to the slow evaporation as well. To accelerate the evaporation phasc, droplets of the NMs/polymer suspension can be added to a hot suspension that is rotating. Using VARTM, the modified epoxy resin was transferred into the fiber layers (VARTM), as described in Figure 2.2.

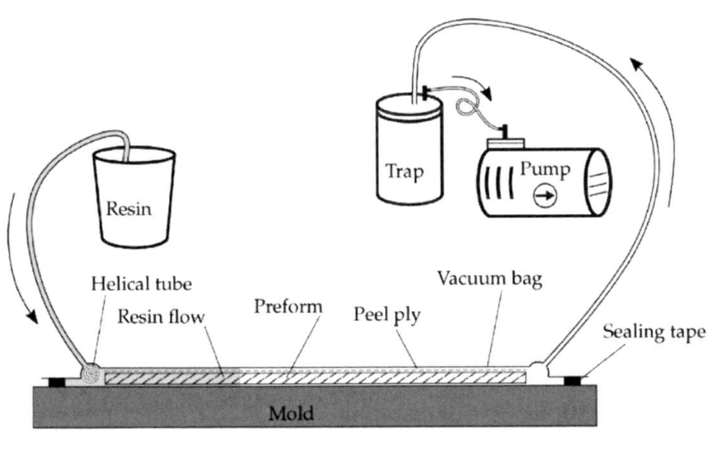

Figure 2.2. VARTM setup (D. Wu et al., 2019).

Hand lay-up is the most common method for making laminated components because it requires fewer facilities. As a result, industrial practices commonly use a variety of low-cost manufacturing procedures to maximize properties while reducing the likelihood of voids and pores. As a result, the next section of this chapter is devoted to a thorough investigation of the properties associated with laminates produced using the hand lay-up approach.

2.2. Analysis of Mechanical Properties

2.2.1. Tensile Strength

The novel material (LPNC) that has been developed is assessed for its ability to withstand tensile stress. Measurement and analysis of the uniaxial tension were conducted by ASTM-D3039 (Figure 2.3). The stress-strain curves acquired from the test were employed to determine the modulus of elasticity and tensile strength (Figure 2.4). Strain measurement is conducted using an MTS810 hydraulic testing machine. The specimen is equipped with an extensometer, and three tests were performed for each weight percentage.

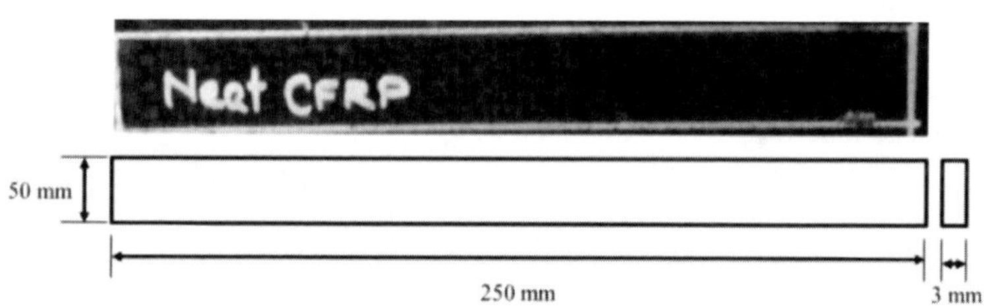

Figure 2.3. Dimension of tensile test specimen (ASTM-D3039).

For reference, neat CF epoxy stress-strain performance is examined for 1, 2, and 3 wt.% GO filler. It was found that neat CF/Epoxy has a tensile strength of 625.79 MPa. Tensile strength goes up by 33.02% when 1% GO is added. The tensile strength of the plain CF/Epoxy composite is enhanced by 20.61% and 19.14% with the addition of 2 wt.% and 3 wt.% GO, respectively. Excessive reinforcing makes LPNC brittle and decreases its tensile strength by agglomerating (Ayatollahi et al., 2011; Ding & Pacek, 2008; F. Li et al., 2015a; Marković et al., 2008). According to the experimental results, the GO concentration should not be too high due to the disadvantage of the stress concentration region in the epoxy (resin) (Rafiee et al., 2019). The tensile strength of developed composites was determined to be 625.79 MPa for neat CF/Epoxy and 934.35 MPa after adding 1 wt.% GO to the epoxy (displayed in Figure 2.5). The young modulus of the pure CF/Epoxy is 8.643 GPa. 10.727 GPa increases the young modulus by the addition of 1 wt.% GO. The inclusion of 2 wt.% GO reduces the modulus to 9.779 GPa. The tensile modulus is minimum at 9.416 GPa with the further incorporation of 3 wt.% GO (displayed in Figure 2.6). The properties may be slightly reduced due to polymer chain motion restriction following high filler loading (Karatas et al., 2019; Prusty et al., 2017a).

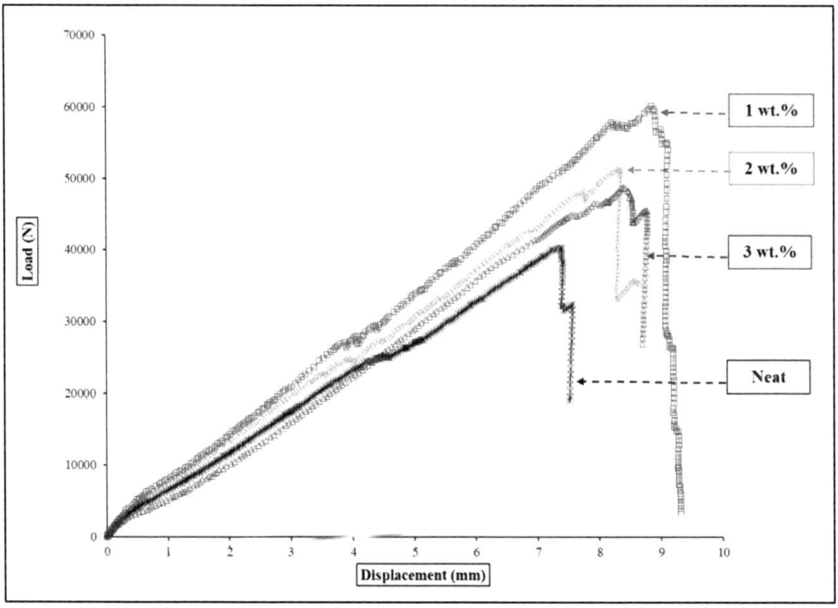

Figure 2.4. Load-displacement curve for tensile properties.

Improvements in the mechanical, electrical, and thermal properties of the polymer matrix were achieved with the incorporation of GO into the epoxy matrix in addition to fiber (Pathak et al., 2016a; Rafiee et al., 2019) (Smith et al., 2019). Based on the literature that has been reported, it is feasible to implement a manual layup processing method to achieve cost-effectiveness (Bauswein et al., 2017; Sałasińska et al., 2022). The enhancement of GO composites is a classic case in point because it typically results in better mechanical properties. As a result, it is crucial to find the optimal combinations of factors to produce the desired nanocomposite properties. The mechanical characteristics of the GO/polymer nanocomposite and the loading of fillers under various circumstances were tested in a few studies. Table 2.1 summarizes the work done by renowned researchers on GO incorporation in the development of polymer composites. The importance of CNMs and polymer properties enhancement for multifunctional products was discussed in this table.

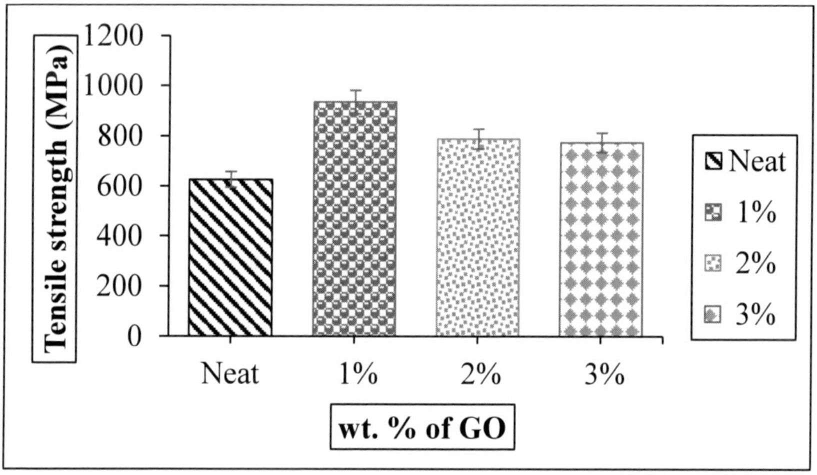

Figure 2.5. Tensile strength (MPa).

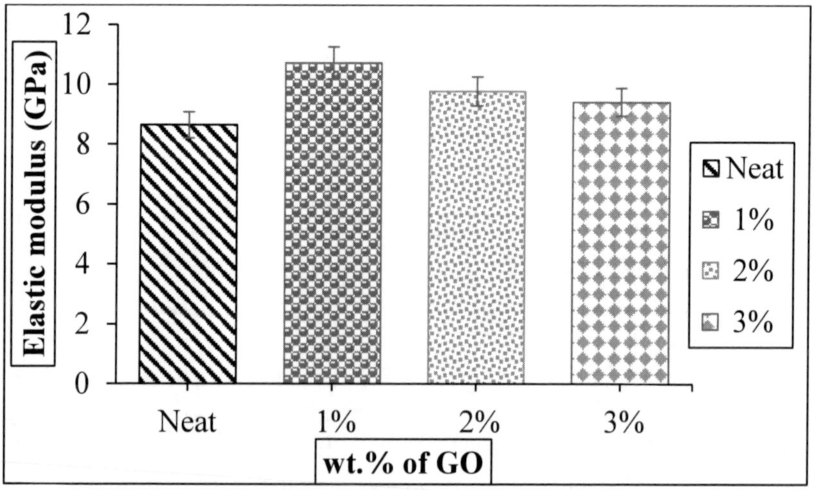

Figure 2.6. Tensile modulus (GPa).

Table 2.1. Young modulus and tensile strength with various CNMs

Nanocomposites	(wt.%)	Tensile properties		Ref.
		Strength (MPa)	Modulus (GPa)	
GNP/MWCNT- EPON-128	0.9/0.1	54.8	3.25	(Yang et al., 2011)
GO/ATP- Epoxy	0.2/1.0	81	3.35	(R. Wang et al., 2013)
GNP/MWCNT- Epoxy	0.26/0.24	68	3.1	(W. Li et al., 2013b)
GO- Polyamide	2.0	179.79	2.89	(Qian et al., 2014)
GO/CF- Epoxy	0.3	737	33.4	(Ashori et al., 2015a)
GO/Short CF- Polyether sulphone	0.5/12.5	119.09	7.79	(F. Li et al., 2015a)
GNP/short CFs- Polypropylene	1.0/15	43.4	1.052	(Ashori et al., 2016)
GO/Short GF- Polyether sulphone	0.5	119	7.3	(Du et al., 2016a)
GNP/CTBN- Epoxy	1/10	68.5	3.78	(F. Wang et al., 2016)
GO/CF- Polyurethane	0.5/1	51.07	4.17	(Jiang et al., 2016)
rGO- Epoxy	2	63	1.95	(L. Chen et al., 2016)
FG/PU- Epoxy	0.8	9.6	0.277	(L. Zhang et al., 2016)
GNP- Epoxy	0.5	88.99	5.1	(Moosa et al., 2016)
GNP- Epoxy	1.0	51.65	1.65	(Zakaria et al., 2017a)
MWCNT- Epoxy	1.0	58.65	1.87	(Zakaria et al., 2017a)
GO/CF- LAPOX C-51	0.2	336	7.3	(Adak, Chhetri, Kim, et al., 2018a)
GO/CF- LAPOX C-5	0.2	454	-	(Adak et al., 2019)
Spray GO/CF- MGS RIM 135	0.5	46.0	9.6	(Vázquez-Moreno et al., 2019)
Spray CNT/CF- MGS RIM 135	0.5	48.9	9.4	(Vázquez-Moreno et al., 2019)
GO/CF EP-100	2	672	-	(Altin et al., 2020)
pGO/CF- EPON 828	0.5	869	6.981	(Shagor et al., 2021)
sGO/CF- EPON 828	0.5	918	7.633	(Shagor et al., 2021)

2.2.2. Flexural Strength

It determines the materials; the flexural properties of the three-point Bending test were carried out on specimens prepared according to ASTM D790 (Figure 2.7), and a 45mm Span length was considered for the same. Universal testing machine MTS810 hydraulic testing machine (Figure 2.8) with a maximum capacity of 100 KN is used for experiments.

It was determined that the flexural strength of the developed composites is 289.95 MPa when pristine CF/epoxy is used. In addition, the flexural strength of the material was enhanced by 442.55, 363.7, and 342.67 MPa after the incorporation of 1, 2, and 3 wt.% GO, as illustrated in Figure 2.9. Stiffness measurements are conducted on the proposed nanocomposite sample using the flexural test. The flexural modulus of the pure CF/Epoxy is 13.807 GPa. With the addition of 1, 2, and 3 wt.% GO, the flexural modules increase by 19.241 GPa, 15.57 GPa, and 11.73 GPa, respectively (presented in Figure 2.10). Table 2.2 summarizes the previous work with different nano fillers for the development of nanocomposites to find the flexural strength.

Figure 2.7. Specimen of flexural test (ASTM D790).

Figure 2.8. Mechanical testing setup.

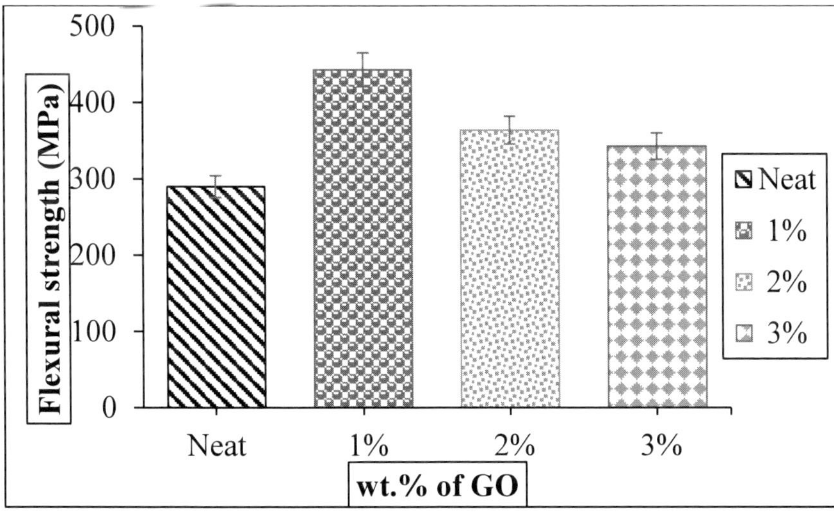

Figure 2.9. Flexural strength (MPa).

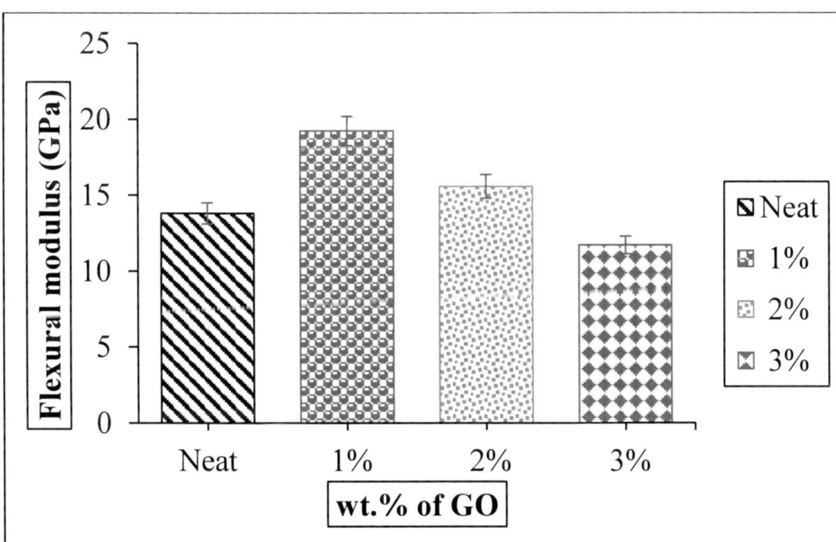

Figure 2.10. Flexural modulus (GPa).

Table 2.2. Flexural strength and flexural modulus with various CNMs

Nanocomposites	(wt.%)	Flexural properties		Ref.
		Strength (MPa)	Modulus (GPa)	
Glass/epoxy/MWCNT	0.75	270	-	(Yip et al., 2011)
Epoxy/nanoclay	1	85.2 ± 2.5	1.6 ± 0.4	(Alamri & Low, 2012)
Cellulose fiber/nanoclay	1	39.13	1.31	(Alamri & Low, 2013)
Unsaturated polyester resin/GNP	0.05	57.6	-	(Swain, 2013)
Epoxy/GO	0.1	130	3.9	(Naebe et al., 2014)
Glass fiber/epoxy/Graphite powder/fly ash	1,1	343.43	1.6678	(Parida et al., 2014)
Carbon fiber/epoxy/GO	0.5	1200	63	(Ashori et al., 2015b)
Polymer/MWCNT	2%	110	2.398	(Babal et al., 2015)
Carbon fiber/epoxy/GO	0.5	182.51 ± 1.51	6.42 ± 0.08	(F. Li et al., 2015b)
Carbon fiber/epoxy/Coated GNP	3	99	11	(Qin et al., 2015a)
Carbon fiber/epoxy/GNP	4	90	11	(Qin et al., 2015b)
Epoxy/MWCNT	0.5	107	6.7	(Vajaiac et al., 2015)
Glass fiber/epoxy/GO	0.5	175	7.6	(Du et al., 2016b)
Carbon fiber/epoxy/GO	0.3	710	35	(Pathak et al., 2016b)
Natural fiber/epoxy/GO	0.05	30	3	(Tshai et al., 2016)
Glass/epoxy/GO	0.5	390	23	(Prusty et al., 2017b)
Carbon fiber/epoxy/GO	0.5	510	35	(Watson et al., 2017)
Polypropylene/MWCNT	10	80	6.250	(Zaghloul, 2017)
Epoxy/MWCNT	1	126.70 ± 2.35	3.35 ± 0.16	(Zakaria et al., 2017b)
Carbon fiber/epoxy/GO	3	492 ± 14	24 ± 3	(Adak, Chhetri, Murmu, et al., 2018a)
Carbon fiber/epoxy/GO	0.2	524	42	(Adak, Chhetri, Kim, et al., 2018b)
Glass fiber/epoxy/GO	2.5	24.523	-	(Arun et al., 2018)
Carbon fiber/epoxy/SWCNT		604.9 ± 29	55.9 ± 2.16	(Burkov & Eremin, 2018)
Glass/epoxy/Graphene	1	250	22	(Umer, 2018)
Carbon fiber/epoxy/MWCNT	0.1	822 ± 27	42.7 ± 1.5	(Xiao et al., 2018)
Carbon fiber/epoxy/MWCNT	1.5	251.1	-	(D. Kumar & Singh, 2019)
Carbon fiber/epoxy/rGO	0.3	340	71	(Jenkins et al., 2019)
Glass fiber/epoxy/GNPs	4%	202.3 ± 5.3	10.0 ± 0.4	(Kunrath et al., 2019)
Epoxy/GO	0.2	82	2.76	(Long et al., 2019)
Epoxy/nanoclay/MWNT	2, 0.02	115.45 ± 0.5	-	(Nguyen et al., 2019)
Glass/epoxy/MWCNT	0.3	332.53	17	(Panchagnula & Kuppan, 2019)
Polypropylene (PP)/Graphite	60	41	7.5	(Rzeczkowski et al., 2019)
Glass fiber/epoxy/GO	0.4	85	7.167	(C. Wu et al., 2019)
Carbon fiber/epoxy/MWCNT	1.5	38.8	-	(Thakur & Singh, 2020)
Glass/epoxy/MWCNT	1	58.4	5.4	(Yildirim et al., 2021)
Jute fiber/epoxy/MWCNT	0.25	13.87	-	(J. Kumar, Kumar, Kumar, et al., 2022b)
Cotton fiber/epoxy/GO	0.3	106.54 ± 4.11	6.996 ± 0.1	(Kamble et al., 2022)
Glass/Carbon/epoxy/MWCNT	3	14.037	-	(Swikker et al., 2022)

There are several uses for materials that combine organic and inorganic polymer groups. The primary goal of these is to develop novel materials with superior characteristics. To manufacture the matrix phase for the fiber-reinforced composite materials, epoxy resin was utilized as a key component. It has found widespread application in adhesives and coatings for laminates due to its excellent mechanical, electrical, abrasion, and chemical reaction resistance (Sałasińska et al., 2022). Superior strength-to-weight ratio, thermal stability, corrosion resistance, stiffness, etc., are only several of the outstanding qualities of polymer composites reinforced by fibers. The manufacturing industry recognizes the importance of these qualities for a wide range of applications, including turbomachinery, sporting goods, airplanes, marine parts, and automobiles (Gang, 2018; Jiménez-Pérez et al., 2017; Sukanchan Palit & Mustansar Hussain, 2018). Its low cost of manufacture, customizable size, surface polishing, safe handling, and other features set it apart from other raw materials for production and transport. In plastics manufacture, fibers capacitate and transfer load to the matrix phase. The percentage of fiber reinforcement mainly determines the mechanical characteristics of a composite. The laminar shear strength and flexural strength qualities are controlled by fiber meshing (D. Chen et al., 2019).

The volumetric interface and filler size are the most critical factors in determining the mechanical performance of composites. As a result, nanofiller has a considerable effect on the mechanical properties of polymer composites in addition to CNMs. Several interface concepts, such as chemical bonding, wetting, mechanical interlocking, and polymer matrix local stiffness, account for the combined effect of the alteration. Elements of the nanofillers family, such as carbon filler, have greatly improved the polymers' mechanical and electrical properties (Domun et al., 2015; Pal et al., 2014). To get the same excellent mechanical qualities with less filler content in the matrix phase, nanofillers are now commonly used in composite manufacture in place of other traditional and nonconventional additives (Ghorai, 2013).

Furthermore, nanofiller alterations in mechanical properties (ILSS, IFSS, and tensile strength) have improved surface modifications for practical use (Adak, Chhetri, Murmu, et al., 2018b; A. Kumar et al., 2019). Thus, composites were filled with fillers having higher mechanical strength and rigidity than polymers (Fan et al., 2010). Nanofillers-modified polymer composites have a better strength/weight ratio than metals and alloys. High filler loads improve mechanical qualities, but they also raise manufacturing costs and make processing more challenging. Being able to improve without significantly increasing the weight of the composite is one of the remarkable benefits of using nanofillers as an addition (Burmistrov et al., 2016; Choudhary & Gupta, 2001; Coleman et al., 2006). In this context, several methods, such as fiber seaming, 3D fabric design, surface fiber, matrix changes, etc., were developed by renowned researchers to improve mechanical aspects. In the case of polymers, the alteration in the matrix phase is significantly more feasible and affordable without compromising other mechanical qualities (Pathak et al., 2016a).

In the literature work of eminent scholars, it has been reported to intend excellent nanofiller materials (Chaudhery & Hussain, 2019; M. S. S. Kumar et al., 2014). Various extensive surveys were proposed to evaluate nanofiller feasibility and their role in manufacturing superior polymer nanocomposites. In this series, by incorporating fullerene (0.1- 1 wt. %) into the resin matrix, Ogasawara et al., (Ogasawara et al., 2009) were able to increase the interlaminar fracture toughness by 60%. Investigated is the extent to which the IFSS changes because of the effects of nanofillers and fiber size (Z. Li et al., 2013). Due to their high specific modulus and strength, CNMs polymer composites have received significant research and innovation focus. It is frequently utilized in a variety of durable items that have high performance and lightweight construction (Stewart, 2009). However, as a laminated structure, CFRP composite laminates are highly susceptible to the initial transmission of cracks and failures as well as fiber/matrix contacts. Therefore, it is vital to develop innovative strategies to improve the CNMs-reinforced polymer interlaminar strength.

Several strategies for improving interfacial characteristics and damage management have been investigated. Similarly, Vaganov et al., (Vaganov et al., 2016) found that incorporating 1% CNT into the epoxy significantly increased the fracture toughness of carbon fiber polymer composites. Carbon fiber polymer matrix composites enhanced with clay material showed a notable improvement in bending capabilities, as validated by the research of Xu et al., (Xu & Hoa, 2008). This research shows that glass fibers with nanostructures perform significantly better in tensile and interface tests. In a related study, Godara et al., (Godara et al., 2009) reported a 90% boost in IFSS intensity in a glass-fiber/epoxy matrix. The glass fibers were directly covered with an epoxy-compatible size based on MWCNT, avoiding the need for any commercial sizing.

Furthermore, the technological difficulty of achieving a homogeneous dispersion of epoxy resins continues. The mechanical properties of nanoparticles can be accessed in CNMs, which are simple to produce at low cost because of the presence of functional groups (rGO, GO platelets, GO sheets, CNO, CNT, etc.). These features improve the dispersion of epoxy matrices and the interfacial interaction of graphene with epoxy (Ning et al., 2015). In polymer composites, several pioneering researchers have demonstrated the feasibility of CNMs (Ashori et al., 2016; Kumari & Kumar, 2018). In the series, Mannov et al., (Mannov et al., 2013) investigated the effects of reduced graphene oxide on the residual compressive behavior of carbon fiber reinforced plastic (CFRP) and glass fiber-reinforced plastic (GFRP) composites Laminates carbon fiber and Glass fiber saw respective increases in compressive residual strength of 35% and 55%. The flexural strength and uniaxial tensile strength of a glass fiber polymer were significantly enhanced after reduced sheets of graphene oxide were added into the material by Yavari et al., (Yavari et al., 2010). CNTs on GNPs in the epoxy matrix are used by Li and Dichiara (W. Li et al., 2013a) to deal with hybrid reinforced composite. The outcome was a rise of 36% in tensile strength and 40% in the Young's modulus of elasticity, respectively.

In the case of polymers, the GO addition results in improved mechanical properties, but the detailed examination of GO in polymers is still in the early stages. To make use of its remarkable qualities for polymer composites, the industrial industry, academia, and research institutions need to pay more attention to their remarkable qualities.

2.2.3. Post-Process Bending and Tensile Tests

Figure 2.11 illustrates the specimen's fracture analysis after the tensile test. During tensile testing, the composite specimen cracked in its middle. Figure 2.12 demonstrates the specimens after the flexural test. Although there is a crack on the fracture surface, it is only on the compressive side of the CFRP material. During the bending, the compressive side of the laminate broke extensively, causing the fiber to collapse catastrophically. Tensile and compressive cracks were much lower in 1 wt.% GO/CF laminates than neat epoxy laminates. The reason for this is the enhanced dispersion of NMs in the matrix. Deterioration, which changes tensile strength and stiffness, is another important issue related to breaking modes. The entire composite structure is severely broken due to insufficient integration at the interphase of two distinct matrices (Yuan et al., 2019).

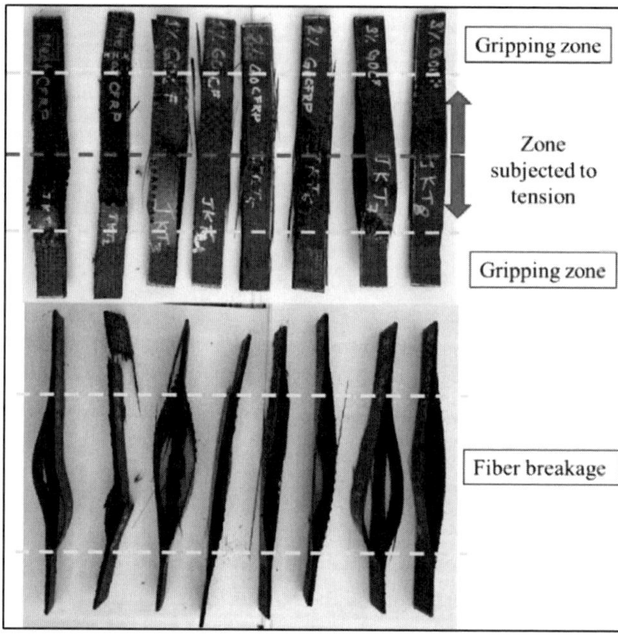

Figure 2.11. Tensile breakages after test.

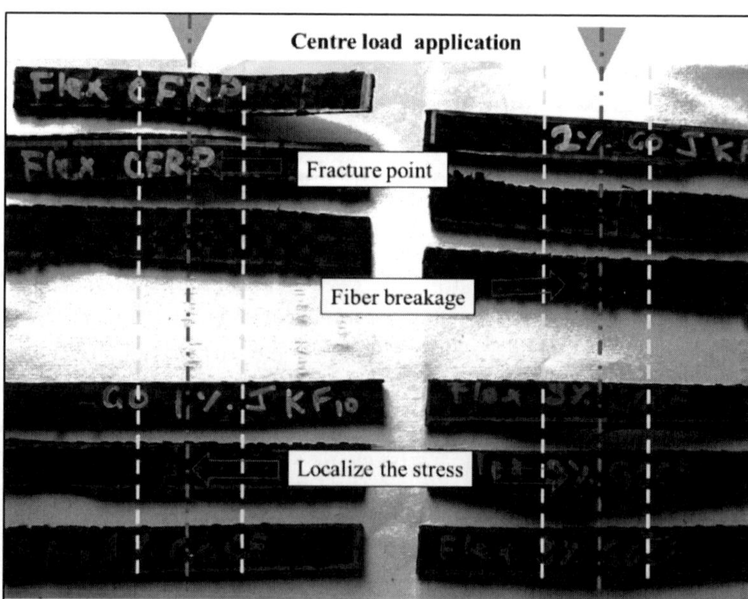

Figure 2.12. Flexural breakages after test.

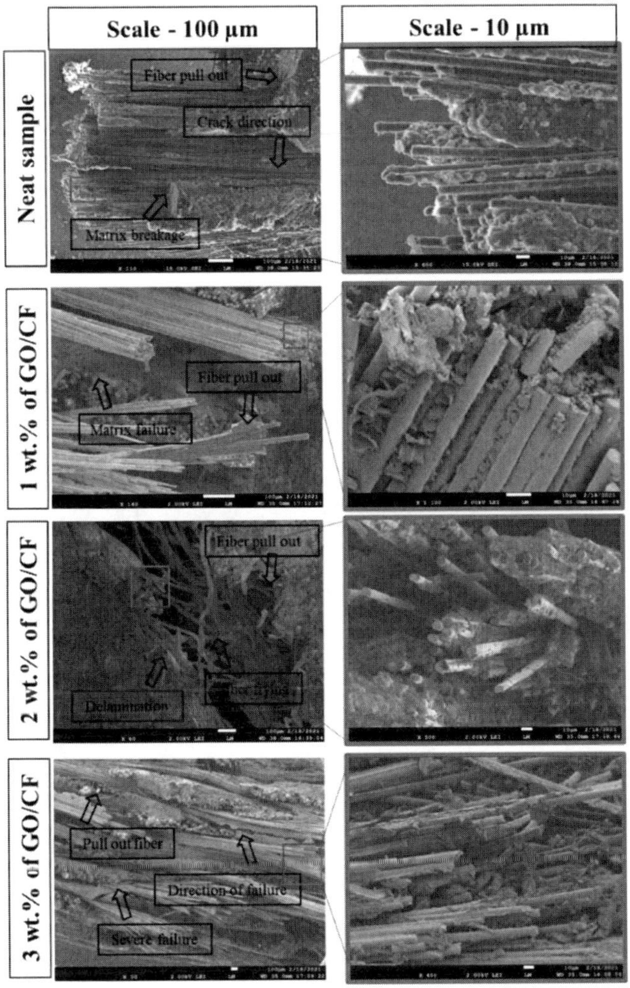

Figure 2.13. Microstructure analysis of tensile test specimen.

On the samples examined, the fracture surface is shown in Figures 2.13 and 2.14. In the case of the neat specimen that was tested, the crack propagation faces exhibited extensive areas of protruding ruptures (ductile). SEM fractography analysis was used to describe the deformation of the GO/CF laminated polymer composite and establish a correlation between the growth of microstructural damage and the fracture process. Figure 2.13 shows clearly the deformations of the polymer, the pulling out of fibers, and the void spaces. Even after matrix cracking has occurred, the matrix continues to provide a mechanism for load transfer due to interfacial shear and fracture interactions. Poor stress transfer between the fiber and matrix, excessive fiber cracking, and aggregation of nanofiller content can contribute to this problem. Figure 2.14 illustrates a brittle failure with fewer fiber pull-outs. Improved bonding and dispersion may be the result of treating fibers with 1% GO. Figure 2.14 depicts an SEM image of the composite specimen's fracture surface after the flexural test. A brittle failure with diminished fiber pull-out was also observed throughout the experiment.

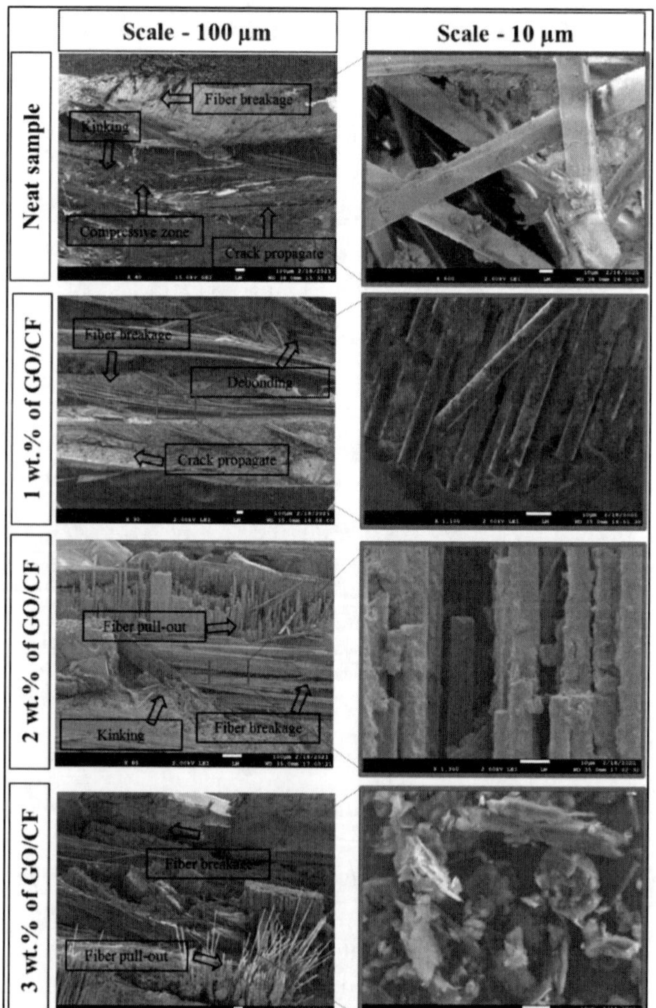

Figure 2.14. Microstructure analysis of flexural test specimen.

2.3. Analysis of Thermal Properties

2.3.1. Dynamic Mechanical Analysis (DMA)

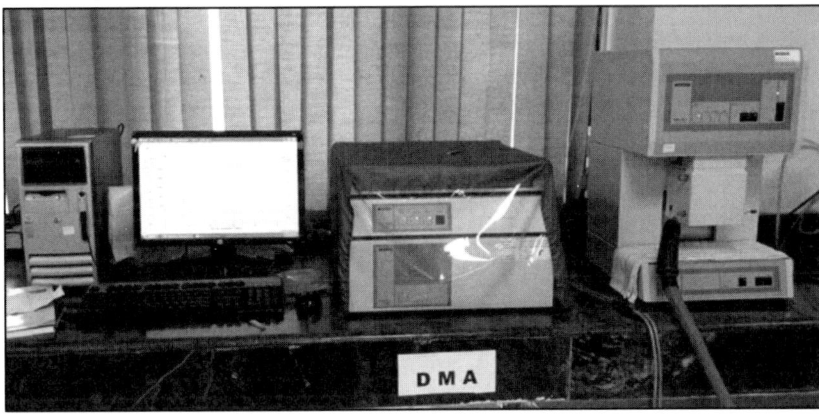

Figure 2.15. DMA setup.

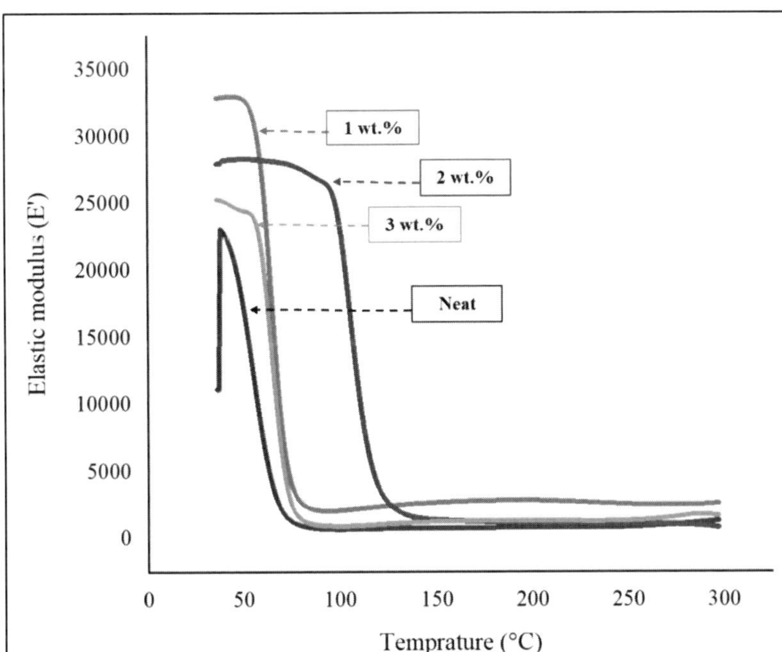

Figure 2.16. GO/CF epoxy nanocomposite storage modulus vs. temperature (°C).

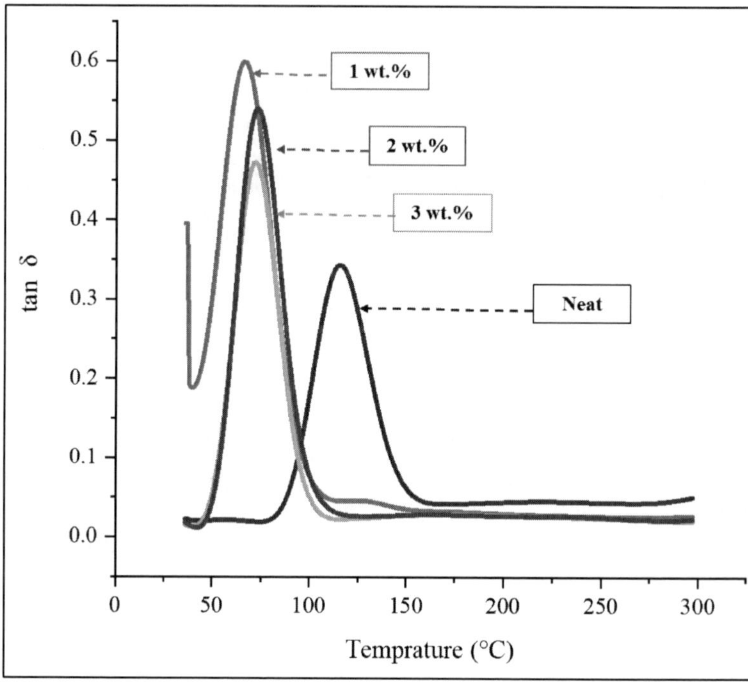

Figure 2.17. GO/CF epoxy nanocomposite tan δ vs temperature (°C).

For characterization, the DMA facility (TGA-55) is depicted in Figure 2.15 with 3-point bending, 5 K/min operating at a temperature of 300°C and a frequency of 1.00 Hz. The function of inclusion of GO nanofiller is proportional to the energy stored per cycle, as illustrated in Figure 2.16 for storage modulus vs. temperature. Figure 2.16 illustrates that the storage modulus of pristine CFRP and GO-modified CFRP composites decreases significantly as the temperature increases because of the epoxy matrix detaching from the reinforcement. It is increased across the entire temperature range of 30 to 100°C by the incorporation of GO in the first segment of the graph, which is below Tg (Adak, Chhetri, Kim, et al., 2018a; Ganguli et al., 2008). In the glassy zone, the storage modulus of the modified CFRP composites increases by 1 weight % of GO. Compared to pristine CFRP, the storage modulus of 1 wt.% GO integrated modified CFRP composites is observed to be 17.47% higher at 35°C. Introducing GO, which restrict polymer-chain mobility, results in an enhanced storage modulus of material (Barkoula et al., 2008; Diani & Gall, 2006).

Additionally, the epoxy polymer's reinforcing effect is limited to certain amounts of GO. Based on DMA evaluation, Figure 2.17 shows the loss factor ($tan\ \delta$) of neat CFRP and its 1 wt.% GO-incorporated modified CFRP nanocomposites. Increased Tg was detected at 1 wt.% GO at 43°C, as indicated by the peak position of tan . In Figure 2.17, GO reinforcement decreased loss factor peak elevation. However, GO nano-filler did not affect peak breadth of tan . A broadened $tan\ \delta$ peak was observed in the nanocomposites on the high-temperature side of the DMA curve. The higher peak factor in nanocomposites indicates less crosslink density and increased diversity, indicating matrix insertion with GO nanoparticles (W. Luo et al., 2017).

2.3.2. Fire Retardant Analysis

Experiments with fire retardants aim to find ways to extinguish flames before they spread and to reduce the rate at which flammable materials burn. Composites that can be used structurally must undergo fire retardant testing to guarantee the safety of people and polymer items in a fire situation. For the fire-retardant test, ASTM D-635 samples were cut to $100 \times 10 \times 10$ mm^3 in accordance with the components being tested. Ignition is accomplished in a horizontal fire setup using a torch for 30 seconds, followed by 30 seconds of specimen self-burning for evaluation purposes.

In cases of horizontal burning, the addition of carbon nanofiller decelerated the onset of the first drip. The MWCNT-0.25 wt. % recommendation took the greatest amount of time before producing its initial drip. The results of the sample testing demonstrated that the nanofiller used in the manufacturing process improved the fire-retardant capabilities of laminate nanocomposites. The effect of nanofiller in improving fire retardancy property can be determined from the improved testing results of manufactured samples in comparison to the neat sample. Synergistic effects can be attributed to the spatial arrangement of particles on the surface of the composite during polymer combustion and ablation. It is believed that the high specific surface area of carbon nanoparticles is responsible for their catalytic actions (Assaedi et al., 2016). More particles mean more restriction sites for the polymer chain, and more complex detachment of the polymer chain means more pyrolysis heat is needed. Uniform dispersed nanoparticles in the polymer network improve the interfacial interaction between the added nanoparticles and the polymer chain, which limits polymer chain mobility (Pathak et al., 2019). This may be attributable to good dispersion of the nanofiller in the epoxy matrix, and it also takes into consideration the important characteristics (Amin et al., 2015; Araby et al., 2021; Liu et al., 2018; Nguyen et al., 2021; Yu & Lau, 2017).

2.4. Analysis of Physical Properties

2.4.1. Water Absorption Analysis

An absorption test was performed on the produced composites in this experimental investigation. The composite materials were cut into specimens with ASTM D570 dimensions of 79.2 × 25.4 × 12.7 mm^3 immersed in fluid (distilled water) for 24 hours, 10 days at room temperature. At regular intervals, each sample was removed from the water and dried with tissue before being weighed with an electronic balance (every 24 hours). The amount of water content (Mt) was determined using the equation below:

$$M_t\% = \frac{W_f - W_i}{W_i} \tag{2.1}$$

Here, W_f, and W_i %, are the final and initial weights of the sample. After several days in the water, the water absorption of all specimens increased steadily (up to 5-6 days). Once immersed for the first time, the rate of water absorption is quick and linear, but once saturation was achieved, it progressively decreases. Compared to nano-filled composites, neat epoxy absorbs more water with fiber. Only a small amount of liquid was absorbed by the nanocomposite during the immersion duration. GO @ 1 wt. % reduced laminate/epoxy water uptake by 44.19%. The specimen improves water adsorption by GO doping and enhancing reinforcement matrix binding. CNMs hydrophobicity and nanocomposite compression densification cause low porosity. The hydrophilic character of liquid in the composite was altered as a result of the synergistic interaction between GO and the densified compact. According to Ghosh et al., (T. Kumar et al., 2014), the addition of graphene to a nanocomposite decreased water absorption and stopped liquid from absorbing into the epoxy resin. The efficiency of the fiber epoxy matrix in preventing water absorption may have been enhanced by the presence of barrier mechanisms that allow for reduced water permeability into the nanocomposite interface. Similar, Alhuthali et al., (Alhuthali et al., 2012) found a similar effect, reporting that decreasing the nanoparticle percentage reduced water intake. Porosity in the material is caused by an increase in the tendency for agglomeration and the formation of a blowing arc with increasing nanoparticle content (Damari et al., 2017; J. Luo et al., 2017; Simulations et al., 2020). As a result, there is more water in the absorbent than in neat composites.

2.5. Analysis of Morphology Properties

Since they affect the acceptable load distribution and durability, the microstructure and metallurgy of substrates, particularly at the surface, are typically required to be thoroughly understood in high-value production. To characterize the newly developed composite based on the engineering properties of the material and the microstructural study, morphological characterization of the composite has been carried out. Additionally examined in this section are the chemical composition, phase, and crystallinity of the fiber composite samples.

2.5.1. X-Ray Diffraction Analysis (XRD)

To examine the X-ray, diffract of the manufactured fiber structural composites Ultima-IV machines were utilized (Computer-controlled alignment of optical devices). The X-ray diffraction method (XRD) is a cutting-edge tool for assessing structural features such as average particle sizes, crystallinity, strain, and crystal defects, as well as phase and crystal orientation. XRD peaks are produced when a beam of monochromatic X-rays is scattered off each pair of lattice planes at a specified angle. The maximal intensity of the lattice planes is determined by the locations of the atomic positions.

X-ray diffraction, commonly abbreviated as XRD, is a non-destructive testing technique used to examine the structure of crystalline substances. Figure 2.18 illustrates the XRD pattern of graphene oxide (GO), with the carbon peak for GO sheets emerging at 10°, which corresponds to a defined d-spacing of 0.8 nm. It has also been reported in an early study by Marcano et al., (Marcano et al., 2010). Figure 2.19 demonstrates the crystal structure and detects the crystalline phases that exist in the material. The XRD analysis is carried out on PAN analytical Empyrean, Cu– Kα, radiation with a ($\lambda = 15.406$

nm), at 40 kV, 30 mA. The 2θ scanned angle is in the range of 10–90° with a step size of 2.0° min^{-1} (Ashori et al., 2015a; F. Li et al., 2015a). The finding demonstrated that the graphite diffraction peak was 26.54°, lower peaks at 43.2°, which confirms the existence of graphite planes (Davoodi et al., 2018; Jeong et al., 2008). The broadening of the peaks can be attributed to the widespread disordering of the graphitic ring's structure due to the smaller size of the GO. The second relatively less intensity peak is observed between 40° and 50° 2θ angle, which confirms the graphene ring in a disordered structure (Pathak et al., 2016a; R. L. Zhang et al., 2016). Due to the existence of a specific oxygen group on the graphite surface, it is oxidized and explored interlaminar spacing (Adak, Chhetri, Kim, et al., 2018a). In Figure 2.19, the wt.% of GO with 1%, 2%, and 3% are XRD trends similar to a slight deviation.

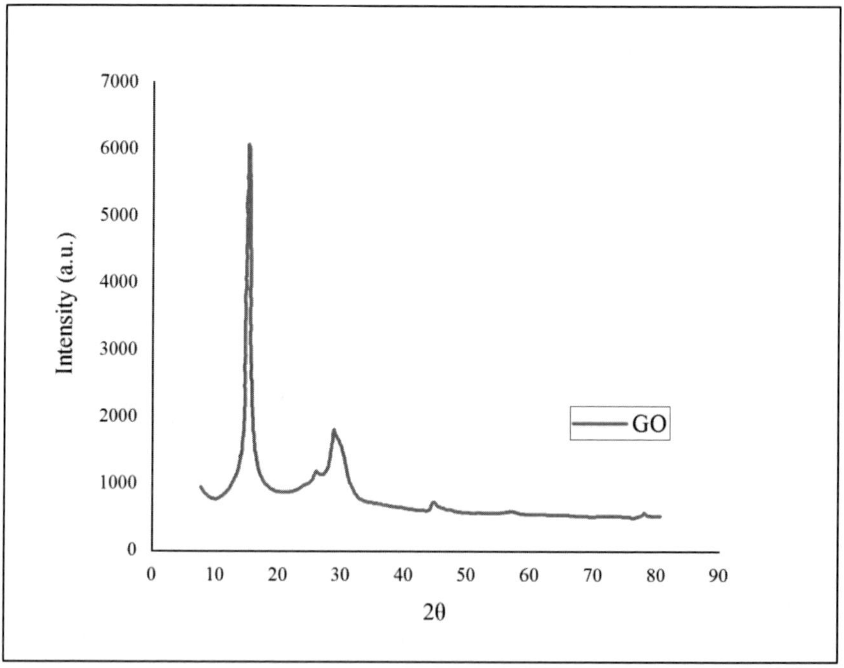

Figure 2.18. XRD pattern of graphene oxide.

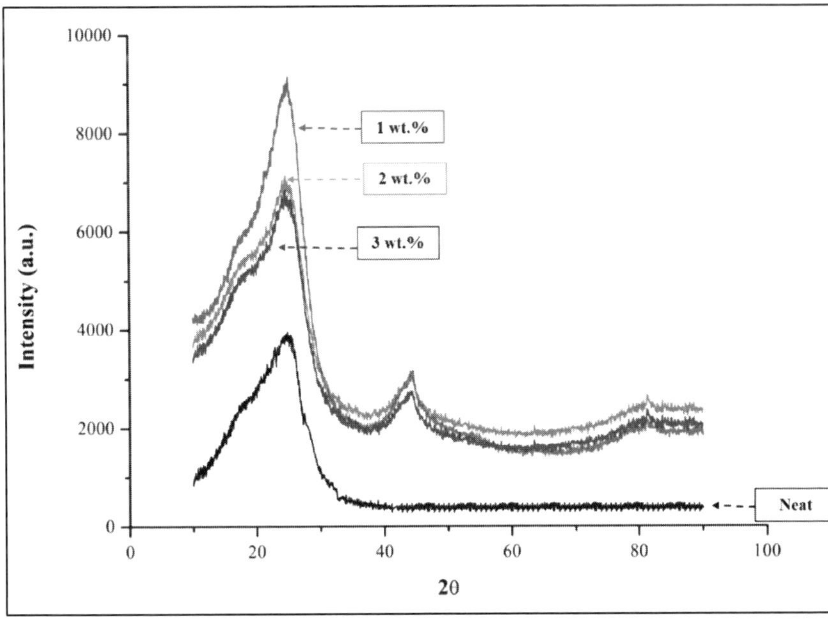

Figure 2.19. XRD pattern of GO/CF nanocomposite.

2.6. Thematic Application of Hybrid Nanocomposites

Figure 2.20 illustrates the fabrication of nanocomposite materials (pipe).

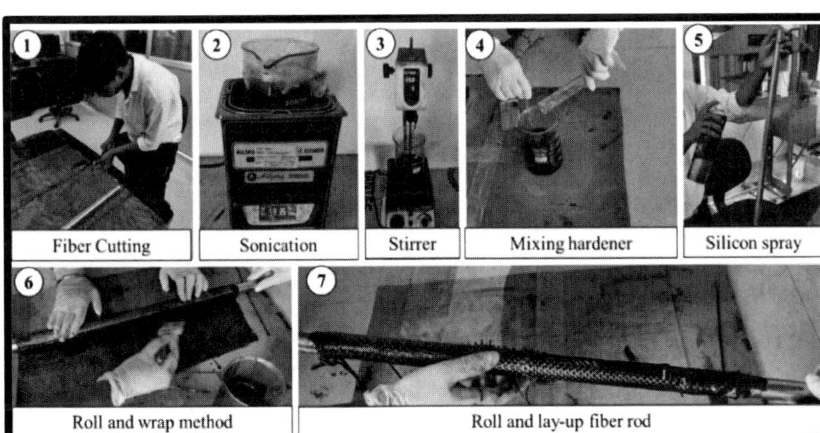

Figure 2.20. Process for development of nanocomposite pipe.

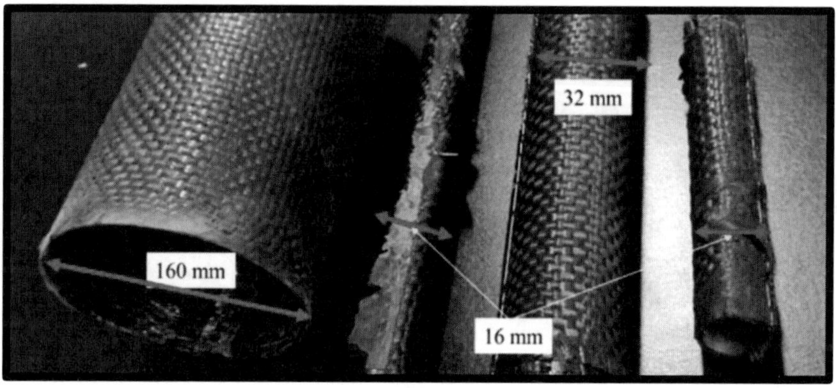

Figure 2.21. Development of nanocomposite pipe.

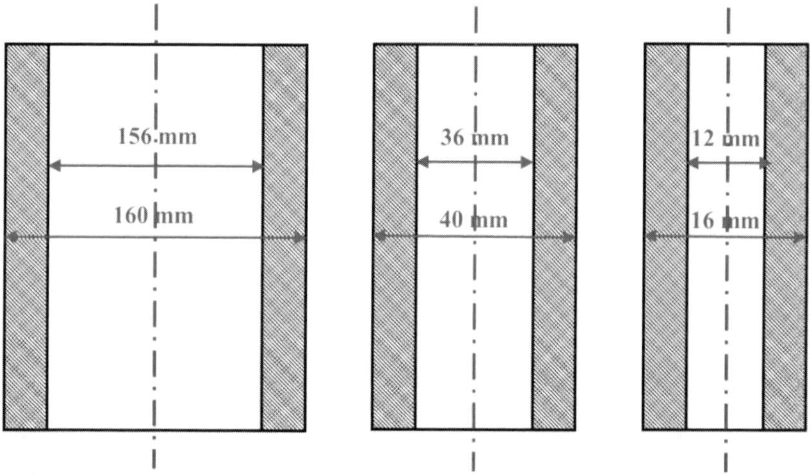

Figure 2.22. Dimension of pipe from GO/CF epoxy nanocomposites.

Figures 2.21 and 2.22 show the developed pipe product from the proposed GO/CF epoxy nanocomposites and their ISO standard dimensions, respectively.

The feasibility and utility for developing end products, as discussed in Chapter 3. The properties of fabricated nanocomposites are specified in Table 2.3. The dissertation's findings show that 1.0 wt.% of graphene oxide was found suitable for improving mechanical properties based on the obtained result. The polymer laminate nanocomposites are used to manufacture automobile components, including covers, mirror housings on various vehicles, door handles, engine covers, belt covers, etc. Also, it is widely used to fabricate automotive, building structure, and under water components (Mohammed et al., 2018). Polymeric composites are implemented enhancements used in the conrod of racing vehicles, storage tank, and sports goods (Myagkov et al., 2014; Nonobe, 2017).

Table 2.3. Mechanical property of the developed GO/CF epoxy nanocomposites

Material	Tensile properties		Reference		Flexural properties		Reference	
	Strength (MPa)	Modulus (MPa)	Li et al., (F. Li et al., 2015a)		Strength (MPa)	Modulus (GPa)	Li et al., (F. Li et al., 2015a)	
Neat	625.79	8.644	106	6.02	289.95	13.807	157.95	4.93
1 wt.% GO	934.35	10.727	117.39	7.92	442.55	19.241	180.62	6.47
2 wt.% GO	788.26	9.779			363.70	15.575		
3 wt.% GO	774.01	9.416			342.67	11.732		

The preliminary investigation based on tensile and flexural strength shows the developed pipe's satisfactory performance. The object produced from initial trials offers good performance in terms of tensile and flexural strength. The initial performance has been satisfactorily endorsed by M/s Jain Fiber Industrial Area Mahaveerpuram, Gorakhnath, Gorakhpur-273015 India for the next phase of assessment and industrial requirements. After some additional characterization and advanced manufacturing steps, it can be endorsed to the polymer manufacturing sector to produce end products and other items.

Conclusion

This chapter emphasizes the progressive characteristics of polymer nanocomposites and the importance of CNMs and fiber in their progression. The fabricated nanocomposites were examined by XRD analysis and FE-SEM image for the uniform dispersal of the nano particle into the epoxy matrix. The mechanical characterization (Tensile & Flexural), thermal properties (DMA), physical properties (fire retardant and water absorption), and morphology study (XRD and Raman spectroscopy) analysis have been done. The following conclusion can be derived based on the acquired results and the discussion regarding the synergistic effect of NMs and their characteristics:

- DMA test findings showed that adding CNMs to epoxy lowered the tan delta curve peak strength. The interfacial adhesion between the reinforcing materials got better, so the curve of the tan delta peak rises.
- With 1% GO, the tensile strength rises by 934.35 MPa, and the modulus rises by 10.727 GPa. With 1% GO, the flexural strength rises by 442.55 MPa, and the modulus rises by 19.241 GPa.
- The neat epoxy laminates had several cracks than the 1% GO/CF laminates, which shows that they failed in different ways.
- The effect of nanofiller in improving fire retardancy property can be determined from the improved testing results of manufactured samples in comparison to the neat sample.
- The efficiency of the fiber epoxy matrix in preventing water absorption may have been enhanced by the presence of barrier mechanisms that allow for reduced water permeability into the nanocomposite interface.
- For the dispersion analyses of the graphene-based nanocomposites, XRD was used to characterize the functionalized GO.

This chapter will provide a novel direction for the polymer manufacturing sectors to utilize the exceptional properties of the proposed materials for better quality and productivity of the end products. This chapter focuses on the manufacturing aspect of polymer nanocomposites modified by Graphene oxide and carbon fiber. The work is initiated from the development phase to end product fabrication through academic laboratory/industry support. The primary phase involves sample development, mechanical testing, and characterization. After product fabrication, the drilling and milling test was performed to evaluate the machinability of the proposed nanocomposite sample. The secondary phase involves examining defects and damages generated during machining of the GO/CF epoxy nanocomposites. These are assessed to identify the application potential for cost-effective and durable product development. In this experimental investigation, the control of varying process constraints plays a significant role in balancing the machining damages. The development of hybrid optimization modules in this work is another exciting feature for the control of process parameters. Due to its superior mechanical properties, including corrosion and moisture resistance, wear resistance, flexible and tensile behaviour, and enhanced strength at a reduced weight, the proposed material has the potential to become one of the most effective alternatives to replace heavyweight and costly engineering materials. As a result, scientists' studying materials concentrate on creating these kinds of popular materials. CNMs could enable material researchers to substitute or supplement carbon fiber-reinforced polymer (CFRP) composites, which are used in aircraft, aerospace, structural, and other high-performance applications. It is highly requisite for an economical machining environment to understand machinability and parametric optimization through a robust statistical tool. The machining databank of LPNC is not aptly investigated in previous studies, and it requires more attention from the manufacturing sector and composite practitioners. Generally, the LPNC are fabricated into the net shape, but the secondary machining process is required to fit and join the fabricated component into the main component assembly.

It can alter the modules' efficiency and finally affect the quality and productivity of the process/product. An attempt has been made to fabricate the end products through the proposed LPNC. This job envisaged the fabrication of end products with the help of industry. The composite which is developed in the academic laboratory has been transformed into an end product through industry-institute interface (III) mode. The preliminary investigation based on tensile and flexural strength tests for the end product, i.e., developed pipes, shows satisfactory performance. The object produced from initial trials offers good performance in terms of tensile and flexural strength. This validates the higher application potential of the proposed nanocomposites.

References

Adak, N. C., Chhetri, S., Kim, N. H., Murmu, N. C., Samanta, P., & Kuila, T. (2018a). Static and Dynamic Mechanical Properties of Graphene Oxide-Incorporated Woven Carbon Fiber/Epoxy Composite. *Journal of Materials Engineering and Performance*, *27*(3), 1138–1147.

Adak, N. C., Chhetri, S., Kim, N. H., Murmu, N. C., Samanta, P., & Kuila, T. (2018b). Static and Dynamic Mechanical Properties of Graphene Oxide-Incorporated Woven Carbon Fiber/Epoxy Composite. *Journal of Materials Engineering and Performance*, *27*(3), 1138–1147.

Adak, N. C., Chhetri, S., Murmu, N. C., Samanta, P., Kuila, T., & Lee, J. H. (2018a). Experimental and numerical investigation on the mechanical characteristics of polyethylenimine functionalized graphene oxide incorporated woven carbon fibre/epoxy composites. *Composites Part B: Engineering*, *156*, 240–251. https://doi.org/10.1016/j.compositesb.2018.08.086.

Adak, N. C., Chhetri, S., Murmu, N. C., Samanta, P., Kuila, T., & Lee, J. H. (2018b). Experimental and numerical investigation on the mechanical characteristics of polyethylenimine functionalized graphene oxide incorporated woven carbon fibre/epoxy composites. *Composites Part B: Engineering*, *156*, 240–251. https://doi.org/10.1016/j.compositesb.2018.08.086.

Adak, N. C., Chhetri, S., Sabarad, S., Roy, H., Murmu, N. C., Samanta, P., & Kuila, T. (2019). Direct observation of micro delamination in graphene oxide incorporated carbon fiber/epoxy composite via in-situ tensile test. *Composites Science and Technology*, *177*, 57–65.

Alamri, H., & Low, I. M. (2012). Effect of water absorption on the mechanical properties of nano-filler reinforced epoxy nanocomposites. *Materials and Design*, *42*, 214–222. https://doi.org/10.1016/j.matdes.2012.05.060.

Alamri, H., & Low, I. M. (2013). Effect of water absorption on the mechanical properties of nanoclay filled recycled cellulose fibre reinforced epoxy hybrid nanocomposites. *Composites Part A: Applied Science and Manufacturing*, *44*, 23–31. https://doi.org/10.1016/j.compositesa.2012.08.026.

Alhuthali, A., Low, I. M., & Dong, C. (2012). Characterisation of the water absorption, mechanical and thermal properties of recycled cellulose fibre reinforced vinyl-ester eco-nanocomposites. *Composites Part B: Engineering*, *43*(7), 2772–2781. https://doi.org/10.1016/j.compositesb.2012.04.038.

Althues, H., Henle, J., & Kaskel, S. (2007). Functional inorganic nanofillers for transparent polymers. *Chemical Society Reviews*, *36*(9), 1454–1465. https://doi.org/10.1039/b608177k.

Altin, Y., Yilmaz, H., Unsal, O. F., & Bedeloglu, A. C. (2020). Graphene oxide modified carbon fiber reinforced epoxy composites. *Journal of Polymer Engineering*, *40*(5), 415–420.

Amin, M. S., El-Gamal, S. M. A., & Hashem, F. S. (2015). Fire resistance and mechanical properties of carbon nanotubes – clay bricks wastes (Homra) composites cement. *Construction and Building Materials*, *98*, 237–249. https://doi.org/10.1016/j.conbuildmat.2015.08.074.

Araby, S., Philips, B., Meng, Q., Ma, J., Laoui, T., & Wang, C. H. (2021). Recent advances in carbon-based nanomaterials for flame retardant polymers and composites. *Composites Part B*, *212*, 108675. https://doi.org/10.1016/j.compositesb.2021.108675.

Armentano, I., Marinucci, L., Dottori, M., Balloni, S., Fortunati, E., Pennacchi, M., Becchetti, E., Locci, P., & Kenny, J. M. (2011). Novel poly(L-lactide) PLLA/SWNTs nanocomposites for biomedical applications: Material characterization and biocompatibility evaluation. *Journal of Biomaterials Science, Polymer Edition*, *22*(4–6), 541–556. https://doi.org/10.1163/092050610X487873.

Artyukhin, A. B., Bakajin, O., Stroeve, P., & Noy, A. (2004). Layer-by-Layer Electrostatic Self-Assembly of Polyelectrolyte Nanoshells on Individual Carbon Nanotube Templates. *Langmuir*, *20*(4), 1442–1448. https://doi.org/10.1021/la035699b.

Arun, G. K., Sreenivas, N., Reddy, K. B., Krishna Reddy, K. S., Shashi Kumar, M. E., & Pramod, R. (2018). Investigation on Mechanical Properties of Graphene Oxide reinforced GFRP. *IOP Conference Series: Materials Science and Engineering*, *310*(1). https://doi.org/10.1088/1757-899X/310/1/012158.

Ashori, A., Menbari, S., & Bahrami, R. (2016). Mechanical and thermo-mechanical properties of short carbon fiber reinforced polypropylene composites using exfoliated graphene nanoplatelets coating. *Journal of Industrial and Engineering Chemistry*, *38*, 37–42.

Ashori, A., Rahmani, H., & Bahrami, R. (2015a). Preparation and characterization of functionalized graphene oxide/carbon fiber/epoxy nanocomposites. *Polymer Testing*, *48*, 82–88.

Ashori, A., Rahmani, H., & Bahrami, R. (2015b). Preparation and characterization of functionalized graphene oxide/carbon fiber/epoxy nanocomposites. *Polymer Testing*, *48*, 82–88.

Assaedi, H., Shaikh, F. U. A., & Low, I. M. (2016). Journal of Asian Ceramic Societies Effect of nano-clay on mechanical and thermal properties of geopolymer. *Integrative Medicine Research*, *4*(1), 19–28. https://doi.org/10.1016/j.jascer.2015.10.004.

Ayatollahi, M. R., Shadlou, S., & Shokrieh, M. M. (2011). Fracture toughness of epoxy/multi-walled carbon nanotube nano-composites under bending and shear loading conditions. *Materials and Design*, *32*(4), 2115–2124.

Babal, A. S., Gupta, R., Singh, B. P., & Dhakate, S. R. (2015). Depression in glass transition temperature of multiwalled carbon nanotubes reinforced polycarbonate composites: Effect of functionalization. *RSC Advances*, *5*(54), 43462–43472. https://doi.org/10.1039/c5ra05825b.

Barkoula, N. M., Alcock, B., Cabrera, N. O., & Peijs, T. (2008). Flame-Retardancy Properties of Intumescent Ammonium Poly(Phosphate) and Mineral Filler Magnesium Hydroxide in Combination with Graphene. *Polymers and Polymer Composites*, *16*(2), 101–113. https://doi.org/10.1002/pc.

Bauswein, Y., Veldenz, L., & Ward, C. (2017). Developing a cost comparison technique for hand layup versus automated fibre placement, and infusion versus out-of-autoclave. *SAMPE Europe Conference 2017*, *June*, 1–8.

Budka, K. C., Deshpande, J. G., Doumi, T. L., Madden, M., & Mew, T. (2010). Smart Grid Communications Equipment: EMI, Safety, and Environmental Compliance Testing Considerations. *Bell Labs Technical Journal*, *15*(2), 205–227.

Burkov, M., & Eremin, A. (2018). Hybrid CFRP/SWCNT Composites with Enhanced Electrical Conductivity and Mechanical Properties. *Journal of Materials Engineering and Performance*, *27*, 5984–5991.

Burmistrov, I., Gorshkov, N., Ilinykh, I., Muratov, D., Kolesnikov, E., Anshin, S., Mazov, I., Issi, J. P., & Kusnezov, D. (2016). Improvement of carbon black based polymer composite electrical conductivity with additions of MWCNT. *Composites Science and Technology*, *129*, 79–85. https://doi.org/10.1016/j.compscitech.2016.03.032.

Chaudhery, S. P., & Hussain, M. (2019). Nanomaterials for Environmental Science: A Recent and Future Perspective. In C. M. Hussain & Ajay Kumar Mishra (Eds.), *Nanotechnology in Environmental Science* (Vol. 2, pp. 1–872). John Wiley & Sons. https://doi.org/10.1039/9781782623625.

Chen, B., & Evans, J. R. G. (2004). Preferential intercalation in polymer-clay nanocomposites. *Journal of Physical Chemistry B*, *108*(39), 14986–14990. https://doi.org/10.1021/jp040312e.

Chen, D., Sun, G., Meng, M., Jin, X., & Li, Q. (2019). Flexural performance and cost efficiency of carbon/basalt/glass hybrid FRP composite laminates. *Thin-Walled Structures*, *142*, 516–531.

Chen, L., Li, W., Liu, Y., & Leng, J. (2016). Nanocomposites of epoxy-based shape memory polymer and thermally reduced graphite oxide: Mechanical, thermal and shape memory characterizations. *Composites Part B: Engineering*, *91*, 75–82.

Choudhary, V., & Gupta, A. (2001). Polymer/Carbon Nanotube Nanocomposites. In S. Yellampalli (Ed.), *Carbon Nanotubes - Polymer Nanocomposites* (Issue July 2015, pp. 65–90). Intech Open. https://doi.org/10.5772/18423.

Coleman, J. N., Khan, U., Blau, W. J., & Gun'ko, Y. K. (2006). Small but strong: A review of the mechanical properties of carbon nanotube-polymer composites. *Carbon*, *44*(9), 1624–1652. https://doi.org/10.1016/j.carbon.2006.02.038.

Damari, S. P., Cullari, L., Nadiv, R., Nir, Y., Laredo, D., & Regev, O. (2017). Graphene-induced enhancement of water vapor barrier in polymer nanocomposites. *Composites Part B*, *134*, 218–224. https://doi.org/10.1016/j.compositesb.2017.09.056.

Davoodi, A. H., Mazinani, S., Sharif, F., & Ranaei-Siadat, S. O. (2018). GO nanosheets localization by morphological study on PLA-GO electrospun nanocomposite nanofibers. *Journal of Polymer Research*, *25*(9), 16–19.

De, S., Patel, A., & Lutkenhaus, J. L. (2020). Layer-by-layer assembly of polymers and anisotropic nanomaterials using spray-based approach. *Journal of Materials Research*, *35*(9), 1163–1172. https://doi.org/10.1557/jmr.2020.44.

Diani, J., & Gall, K. (2006). Finite Strain 3D Thermoviscoelastic Constitutive Model. *Society*, *46*(4), 1–10.

Ding, P., & Pacek, A. W. (2008). De-agglomeration of goethite nano-particles using ultrasonic comminution device. *Powder Technology*, *187*(1), 1–10.

Domun, N., Hadavinia, H., Zhang, T., Sainsbury, T., Liaghat, G. H., & Vahid, S. (2015). Improving the fracture toughness and the strength of epoxy using nanomaterials-a review of the current status. *Nanoscale*, *7*(23), 10294–10329. https://doi.org/10.1039/C5NR01354B.

Du, S. Sen, Li, F., Xiao, H. M., Li, Y. Q., Hu, N., & Fu, S. Y. (2016a). Tensile and flexural properties of graphene oxide coated-short glass fiber reinforced polyethersulfone composites. *Composites Part B: Engineering*, *99*, 407–415.

Du, S. Sen, Li, F., Xiao, H. M., Li, Y. Q., Hu, N., & Fu, S. Y. (2016b). Tensile and flexural properties of graphene oxide coated-short glass fiber reinforced polyethersulfone composites. *Composites Part B: Engineering*, *99*, 407–415.

Fan, H., Wang, L., Zhao, K., Li, N., Shi, Z., Ge, Z., & Jin, Z. (2010). Fabrication, mechanical properties, and biocompatibility of graphene-reinforced chitosan composites. *Biomacromolecules*, *11*(9), 2345–2351. https://doi.org/10.1021/bm100470q.

Gang, D. (2018). The effect of surface treatment of CF and graphene oxide on the mechanical properties of PI composite. *Journal of Thermoplastic Composite Materials*, *31*(9), 1219–1231. https://doi.org/10.1177/0892705717734606.

Ganguli, S., Roy, A. K., & Anderson, D. P. (2008). Improved thermal conductivity for chemically functionalized exfoliated graphite/epoxy composites. *Carbon*, *46*(5), 806–817.

Ghorai, S. (2013). Chemical, physical and mechanical properties of nanomaterials and its applications [University of Iowa]. In *University of Iowa*. https://doi.org/10.17077/etd.5r1xf0y8.

Godara, A., Mezzo, L., Luizi, F., Warrier, A., Lomov, S. V., van Vuure, A. W., Gorbatikh, L., Moldenaers, P., & Verpoest, I. (2009). Influence of carbon nanotube reinforcement on the processing and the mechanical behaviour of carbon fiber/epoxy composites. *Carbon*, *47*(12), 2914–2923.

Gromov, A., Dittmer, S., Svensson, J., Nerushev, O. A., Perez-García, S. A., Licea-Jiménez, L., Rychwalski, R., & Campbell, E. E. B. (2005). Covalent amino-functionalisation of single-wall carbon nanotubes. *Journal of Materials Chemistry*, *15*(32), 3334–3339.

Grothe, J., Kaskel, S., & Leuteritz, A. (2012). Nanocomposites and Hybrid Materials. In *Polymer Science: a Comprehensive Reference: Volume 1-10* (Vols. 1–10). Elsevier B.V. https://doi.org/10.1016/B978-0-444-53349-4.00206-5.

Hanemann, T., & Szabó, D. V. (2010). Polymer-nanoparticle composites: From synthesis to modern applications. *Materials*, *3*(6), 3468–3517. https://doi.org/10.3390/ma3063468.

Jenkins, P., Siddique, S., Khan, S., Usman, A., Starost, K., MacPherson, A., Bari, P., Mishra, S., & Njuguna, J. (2019). Influence of Reduced Graphene Oxide on Epoxy/Carbon Fiber-Reinforced Hybrid Composite: Flexural and Shear Properties under Varying Temperature Conditions. *Advanced Engineering Materials*, *21*(6), 1–14. https://doi.org/10.1002/adem.201800614.

Jeong, H. K., Yun, P. L., Lahaye, R. J. W. E., Park, M. H., Kay, H. A., Ick, J. K., Yang, C. W., Chong, Y. P., Ruoff, R. S., & Young, H. L. (2008). Evidence of graphitic AB stacking order of graphite oxides. *Journal of the American Chemical Society*, *130*(4), 1362–1366.

Jiang, S., Li, Q., Wang, J., He, Z., Zhao, Y., & Kang, M. (2016). Multiscale graphene oxide-carbon fiber reinforcements for advanced polyurethane composites. *Composites Part A: Applied Science and Manufacturing*, *87*, 1–9.

Jiménez-Pérez, V. M., Kharissova, O. V., & Flores, B. M. M. (2017). Environmental Applications of Iron-Containing Nanomaterials: Synthetic Routes, Structures, Compositions and Properties. In C. M. Hussain & B. Kharisov (Eds.), *Advanced Environmental Analysis: Applications of Nanomaterials* (Vol. 1, pp. 193–220). The Royal Society of Chemistry.

Kamble, Z., Behera, B. K., Kimura, T., & Haruhiro, I. (2022). Development and characterization of thermoset nanocomposites reinforced with cotton fibres recovered from textile waste. *Journal of Industrial Textiles*, *51*(2), 2026S-2052S. https://doi.org/10.1177/1528083720913535.

Karatas, E., Gul, O., Karsli, N. G., & Yilmaz, T. (2019). Synergetic effect of graphene nanoplatelet, carbon fiber and coupling agent addition on the tribological, mechanical and thermal properties of polyamide 6,6 composites. *Composites Part B: Engineering*, *163*, 730–739.

Kumar, A., Sharma, K., & Dixit, A. R. (2019). A review of the mechanical and thermal properties of graphene and its hybrid polymer nanocomposites for structural applications. *Journal of Materials Science*, *54*(8), 5992–6026. https://doi.org/10.1007/s10853-018-03244-3.

Kumar, D., & Singh, K. K. (2019). Investigation of delamination and surface quality of machined holes in drilling of multiwalled carbon nanotube doped epoxy/carbon fiber reinforced polymer nanocomposite. *Proceedings of the Institution of Mechanical Engineers, Part L: Journal of Materials: Design and Applications*, *233*(4), 647–663.

Kumar, J., Jaiswal, B., Kumar, K., Kumar, K., & Verma, R. K. (2023). Reuse of waste carpet into a sustainable composite: A case study on recycling approach. *Sustainable Materials and Manufacturing Technologies*, 230–243. https://doi.org/10.1201/9781003291961-17.

Kumar, J., Kesarwani, S., Kharwar, P. K., Jackson, M. J., & Verma, R. K. (2023). Mechanical performance and drilling machinability evaluation of carbon nano onions (CNOs) reinforced polymer nanocomposites. *International Journal on Interactive Design and Manufacturing*, *17*(1), 169–186. https://doi.org/10.1007/s12008-022-01160-0.

Kumar, J., Kumar, K., Jaiswal, B., Kumar, K., & Verma, R. K. (2022). Investigation on the physio-mechanical properties of carpet waste polymer composites incorporated with multi-wall carbon nanotube (MWCNT). *Journal of the Textile Institute*, *0*(0), 1–10. https://doi.org/10.1080/00405000.2022.2062860.

Kumar, J., Kumar, K. K., Kumar, K. K., Jaiswal, B., & Verma, R. K. (2022a). Development of waste carpet (jute) and Multi-wall carbon nanotube incorporated epoxy composites for lightweight applications. *Progress in Rubber Plastics and Recycling Technology*, *38*(3), 247–263. https://doi.org/10.1177/14777606221110252.

Kumar, J., Kumar, K. K., Kumar, K. K., Jaiswal, B., & Verma, R. K. (2022b). Development of waste carpet (jute) and Multi-wall carbon nanotube incorporated epoxy composites for lightweight applications. *Progress in Rubber Plastics and Recycling Technology*, *38*(3), 247–263. https://doi.org/10.1177/14777606221110252.

Kumar, J., Kumar Verma, R., & Debnath, K. (2020). A new approach to control the delamination and thrust force during drilling of polymer nanocomposites reinforced by graphene oxide/carbon fiber. *Composite Structures*, *253*(August), 112786. https://doi.org/10.1016/j.compstruct.2020.112786.

Kumar, J., Verma, R. K., & Mondal, A. K. (2020). Predictive modeling and machining performance optimization during drilling of polymer nanocomposites reinforced by graphene oxide/carbon fiber. *Archive of Mechanical Engineering*, *67*(2), 229–258. https://doi.org/10.24425/ame.2020.131692.

Kumar, J., Verma, R. K., Mondal, A. K., & Singh, V. K. (2021). A hybrid optimization technique to control the machining performance of graphene/carbon/polymer (epoxy) nanocomposites. *Polymers and Polymer Composites*, *29*(9), S1168–S1180. https://doi.org/10.1177/09673911211046789.

Kumar, M. S. S., Sundara Raju, N. M., Sampath, P. S., & Jayakumari, L. S. (2014). Effects of nanomaterials on polymer composites - An expatiate view. *Reviews on Advanced Materials Science*, *38*(1), 40–54.

Kumar, T., Gope, S., Mondal, D., Bhowmik, B., Rahaman, M., Maity, D., Roy, I., Sarkar, G., Sadhukhan, S., Rana, D., Chakraborty, M., & Chattopadhyay, D. (2014). Assessment of morphology and property of graphene oxide-hydroxypropylmethylcellulose nanocomposite films. *International Journal of Biological Macromolecules*, *66*, 338–345. https://doi.org/10.1016/j.ijbiomac.2014.02.054.

Kumari, N., & Kumar, K. (2018). Mechanical behaviour of graphene and carbon fibre reinforced epoxy based hybrid nanocomposites for orthotic callipers. *Journal of Experimental Nanoscience*, *13*, S14–S23.

Kunrath, K., Kerche, E. F., Rezende, M. C., & Amico, S. C. (2019). Mechanical, electrical, and electromagnetic properties of hybrid graphene/glass fiber/epoxy composite. *Polymers and Polymer Composites*, *27*(5), 1–6. https://doi.org/10.1177%2F0967391119828559.

Landry, C. J. T., Coltrain, B. K., Wesson, J. A., Zumbulyadis, N., & Lippert, J. L. (1992). In situ polymerization of tetraethoxysilane in polymers: chemical nature of the interactions. *Polymer*, *33*(7), 1496–1506. https://doi.org/10.1016/0032-3861(92)90127-I.

Li, F., Liu, Y., Qu, C. B., Xiao, H. M., Hua, Y., Sui, G. X., & Fu, S. Y. (2015a). Enhanced mechanical properties of short carbon fiber reinforced polyethersulfone composites by graphene oxide coating. *Polymer*, *59*, 155–165. https://doi.org/10.1016/j.polymer.2014.12.067.

Li, F., Liu, Y., Qu, C. B., Xiao, H. M., Hua, Y., Sui, G. X., & Fu, S. Y. (2015b). Enhanced mechanical properties of short carbon fiber reinforced polyethersulfone composites by graphene oxide coating. *Polymer*, *59*, 155–165. https://doi.org/10.1016/j.polymer.2014.12.067.

Li, W., Dichiara, A., & Bai, J. (2013a). Carbon nanotube – graphene nanoplatelet hybrids as high-performance multifunctional reinforcements in epoxy composites. *COMPOSITES SCIENCE AND TECHNOLOGY*, *74*, 221–227.

Li, W., Dichiara, A., & Bai, J. (2013b). Carbon nanotube-graphene nanoplatelet hybrids as high-performance multifunctional reinforcements in epoxy composites. *Composites Science and Technology*, *74*, 221–227.

Li, Z., Wang, J., Liu, X., Liu, S., Ou, J., & Yang, S. (2011). Electrostatic layer-by-layer self-assembly multilayer films based on graphene and manganese dioxide sheets as novel electrode materials for supercapacitors. *Journal of Materials Chemistry*, *21*(10), 3397–3403. https://doi.org/10.1039/c0jm02650f.

Li, Z., Wang, R., Young, R. J., Deng, L., Yang, F., Hao, L., Jiao, W., & Liu, W. (2013). Control of the functionality of graphene oxide for its application in epoxy nanocomposites. *Polymer*, *54*(23), 6437–6446.

Lin, C., Wang, Y., Lai, Y., Yang, W., Jiao, F., Zhang, H., Ye, S., & Zhang, Q. (2011). Incorporation of carboxylation multiwalled carbon nanotubes into biodegradable poly(lactic-co-glycolic acid) for bone tissue engineering. *Colloids and Surfaces B: Biointerfaces*, *83*(2), 367–375. https://doi.org/10.1016/j.colsurfb.2010.12.011.

Liu, Z., Li, Z., Yang, Y., Zhang, Y., Wen, X., Li, N., & Fu, C. (2018). A Geometry Effect of Carbon Nanomaterials on Flame Retardancy and Mechanical Properties of Ethylene-Vinyl Acetate / Magnesium Hydroxide Composites. *Polymers*, *10*, 1028 (1-14). https://doi.org/10.3390/polym10091028.

Long, J. P., Li, S. X., Liang, B., & Wang, Z. G. (2019). Investigation of thermal behaviour and mechanical property of the functionalised graphene oxide/epoxy resin nanocomposites. *Plastics, Rubber and Composites*, *48*(3), 127–136. https://doi.org/10.1080/14658011.2019.1577027.

Luo, J., Yang, S., Lei, L., Zhao, J., & Tong, Z. (2017). Composites : Part A Toughening , synergistic fire retardation and water resistance of polydimethylsiloxane grafted graphene oxide to epoxy nanocomposites with trace phosphorus. *Composites Part A*, *100*, 275–284. https://doi.org/10.1016/j.compositesa.2017.05.019.

Luo, W., Zhang, B., Zou, H., Liang, M., & Chen, Y. (2017). Enhanced interfacial adhesion between polypropylene and carbon fiber by graphene oxide/polyethyleneimine coating. *Journal of Industrial and Engineering Chemistry*, *51*, 129–139.

Mannov, E., Schmutzler, H., Chandrasekaran, S., Viets, C., Buschhorn, S., Tölle, F., Mülhaupt, R., & Schulte, K. (2013). Improvement of compressive strength after impact in fibre reinforced polymer composites by matrix modification with thermally reduced graphene oxide. *Composites Science and Technology*, *87*, 36–41.

Marcano, D. C., Kosynkin, D. V., Berlin, J. M., Sinitskii, A., Sun, Z., Slesarev, A., Alemany, L. B., Lu, W., & Tour, J. M. (2010). Improved synthesis of graphene oxide. *ACS Nano*, *4*(8), 4806–4814.

Marković, S., Mitrić, M., Starčević, G., & Uskoković, D. (2008). Ultrasonic de-agglomeration of barium titanate powder. *Ultrasonics Sonochemistry*, *15*(1), 16–20.

Mcshane, M. J., & Lvov, Y. M. (2014). Electrostatic Self-Assembly : Layer-by-Layer Layer-by-Layer Electrostatic Self-Assembly. *Dekker Encyclopedia of Nanoscience and Nanotechnology*, *June*. https://doi.org/10.1081/E-ENN-120013616.

Mohammed, A., Al-Saadi, N. T. K., & Al-Mahaidi, R. (2018). Assessing the contribution of the CFRP strip of bearing the applied load using near-surface mounted strengthening technique with innovative high-strength self-compacting cementitious adhesive (IHSSC-CA). *Polymers*, *10*(1), 66 (1-14).

Moosa, A., Ramazani, A., & Nabil, M. (2016). Mechanical and Electrical Properties of Graphene Nanoplates and Carbon-Nanotubes Hybrid Epoxy Nanocomposites. *American Journal of Materials Science*, *6*(6), 157–165.

Myagkov, L. L., Mahkamov, K., Chainov, N. D., & Makhkamova, I. (2014). Advanced and conventional internal combustion engine materials. In R. Folkson (Ed.), *Alternative Fuels and Advanced Vehicle Technologies for Improved Environmental Performance: Towards Zero Carbon Transportation* (pp. 370–392). Woodhead Publishing Limited. https://doi.org/10.1533/9780857097422.2.370.

Naebe, M., Wang, J., Amini, A., Khayyam, H., Hameed, N., Li, L. H., Chen, Y., & Fox, B. (2014). Mechanical Property and Structure of Covalent Functionalised Graphene/Epoxy Nanocomposites. *Scientific Reports*, *4*, 4375 (1-7). https://doi.org/10.1038/srep04375.

Nguyen, T. A., Nguyen, Q. T., & Bach, T. P. (2019). Mechanical Properties and Flame Retardancy of Epoxy Resin/Nanoclay/Multiwalled Carbon Nanotube Nanocomposites. *Journal of Chemistry*, *2019*, 3105205 (1-9). https://doi.org/10.1155/2019/3105205.

Nguyen, T. A., Thu, T., & Bui, T. (2021). Effects of Hybrid Graphene Oxide with Multiwalled Carbon Nanotubes and Nanoclay on the Mechanical Properties and Fire Resistance of Epoxy Nanocomposite. *Journal of Nanomaterials*, *2021*, 2862426 (1-7). https://doi.org/10.1155/2021/2862426.

Ning, H., Li, J., Hu, N., Yan, C., Liu, Y., Wu, L., Liu, F., & Zhang, J. (2015). Interlaminar mechanical properties of carbon fiber reinforced plastic laminates modified with graphene oxide interleaf. *Carbon, 91*, 224–233. https://doi.org/10.1016/j.carbon.2015.04.054.

Nonobe, Y. (2017). Development of the fuel cell vehicle mirai. *IEEJ Transactions on Electrical and Electronic Engineering*, *12*(1), 5–9. https://doi.org/10.1002/tee.22328.

Ogasawara, T., Ishida, Y., & Kasai, T. (2009). Mechanical properties of carbon fiber/fullerene-dispersed epoxy composites. *Composites Science and Technology*, *69*(11–12), 2002–2007.

Pal, H., Sharma, V., & Sharma, M. (2014). Influence of functionalization on mechanical and electrical properties of carbon nanotube-based silver composites. *Philosophical Magazine*, *94*(13), 1478–1492. https://doi.org/10.1080/14786435.2014.892221.

Panchagnula, K. K., & Kuppan, P. (2019). Improvement in the mechanical properties of neat GFRPs with multi-walled CNTs. *Journal of Materials Research and Technology*, *8*(1), 366–376. https://doi.org/10.1016/j.jmrt.2018.02.009.

Pande, S., Chaudhary, A., Patel, D., Singh, B. P., & Mathur, R. B. (2014). Mechanical and electrical properties of multiwall carbon nanotube/polycarbonate composites for electrostatic discharge and electromagnetic interference shielding applications. *RSC Advances*, *4*(27), 13839–13849.

Parida, A. kumar, Das, R., Sahoo, A. K., & Routara, B. C. (2014). Optimization of Cutting Parameters for Surface Roughness in Machining of gfrp Composites with Graphite/fly Ash Filler. *Procedia Materials Science*, *6*, 1533–1538.

Pathak, A. K., Borah, M., Gupta, A., Yokozeki, T., & Dhakate, S. R. (2016a). Improved mechanical properties of carbon fiber/graphene oxide-epoxy hybrid composites. *Composites Science and Technology*, *135*, 28–38. https://doi.org/10.1016/j.compscitech.2016.09.007.

Pathak, A. K., Borah, M., Gupta, A., Yokozeki, T., & Dhakate, S. R. (2016b). Improved mechanical properties of carbon fiber/graphene oxide-epoxy hybrid composites. *Composites Science and Technology*, *135*, 28–38. https://doi.org/10.1016/j.compscitech.2016.09.007.

Pathak, A. K., Garg, H., Singh, M., Yokozeki, T., & Dhakate, S. R. (2019). Enhanced interfacial properties of graphene oxide incorporated carbon fiber reinforced epoxy nanocomposite: a systematic thermal properties investigation. *Journal of Polymer Research*, *26*(2), 23 (1-23). https://doi.org/10.1007/s10965-018-1668-2.

Prusty, R. K., Ghosh, S. K., Rathore, D. K., & Ray, B. C. (2017a). Reinforcement effect of graphene oxide in glass fibre/epoxy composites at in-situ elevated temperature environments: An emphasis on graphene oxide content. *Composites Part A: Applied Science and Manufacturing*, *95*, 40–53.

Prusty, R. K., Ghosh, S. K., Rathore, D. K., & Ray, B. C. (2017b). Reinforcement effect of graphene oxide in glass fibre/epoxy composites at in-situ elevated temperature environments: An emphasis on graphene oxide content. *Composites Part A: Applied Science and Manufacturing*, *95*, 40–53.

Qian, Y., Lan, Y., Xu, J., Ye, F., & Dai, S. (2014). Fabrication of polyimide-based nanocomposites containing functionalized graphene oxide nanosheets by in-situ polymerization and their properties. *Applied Surface Science*, *314*, 991–999.

Qin, W., Vautard, F., Drzal, L. T., & Yu, J. (2015a). Mechanical and electrical properties of carbon fiber composites with incorporation of graphene nanoplatelets at the fiber-matrix interphase. *Composites Part B: Engineering*, *69*, 335–341. https://doi.org/10.1016/j.compositesb.2014.10.014.

Qin, W., Vautard, F., Drzal, L. T., & Yu, J. (2015b). Mechanical and electrical properties of carbon fiber composites with incorporation of graphene nanoplatelets at the fiber-matrix interphase. *Composites Part B: Engineering*, *69*, 335–341.

Rafiee, M., Nitzsche, F., Laliberte, J., Thibault, J., & Labrosse, M. R. (2019). Simultaneous reinforcement of matrix and fibers for enhancement of mechanical properties of graphene-modified laminated composites. *Polymer Composites*, *40*(S2), E1732–E1745. https://doi.org/10.1002/pc.25137.

Rzeczkowski, P., Krause, B., & Pötschke, P. (2019). Characterization of highly filled PP/graphite composites for adhesive joining in fuel cell applications. *Polymers*, *11*(3), 462. https://doi.org/10.3390/polym11030462.

Sałasińska, K., Cabulis, P., Kirpluks, M., Kovalovs, A., Kozikowski, P., Barczewski, M., Celiński, M., Mizera, K., Gałecka, M., Skukis, E., Kalnins, K., Cabulis, U., & Boczkowska, A. (2022). The Effect of Manufacture Process on Mechanical Properties and Burning Behavior of Epoxy-Based Hybrid Composites. *Materials*, *15*(1), 1–15. https://doi.org/10.3390/ma15010301.

Shagor, R. M. R., Abedin, F., & Asmatulu, R. (2021). Mechanical and thermal properties of carbon fiber reinforced composite with silanized graphene as nano-inclusions. *Journal of Composite Materials*, *55*(5), 597–608. https://doi.org/10.1177/0021998320953183.

Simulations, N., Slusarek, J., Krause, P., Szymanowska-gwizdz, A., Ohmi, T., & Nakamura, M. (2020). Understanding the Role of Graphene on the Moisture Absorption in Graphene / Polymer Nanocomposites Understanding the Role of Graphene on the Moisture Absorption in Graphene / Polymer Nanocomposites. *IOP Conf. Series: Materials Science and Engineering*, 012006. https://doi.org/10.1088/1757-899X/1024/1/012006.

Smith, A. T., LaChance, A. M., Zeng, S., Liu, B., & Sun, L. (2019). Synthesis, properties, and applications of graphene oxide/reduced graphene oxide and their nanocomposites. *Nano Materials Science*, *1*(1), 31–47. https://doi.org/10.1016/j.nanoms.2019.02.004.

Srivastava, S., & Kotov, N. A. (2008). Composite Layer-by-Layer (LBL) assembly with inorganic nanoparticles and nanowires. *Accounts of Chemical Research*, *41*(12), 1831–1841. https://doi.org/10.1021/ar8001377.

Stewart, R. (2009). Lightweighting the automotive market. *Reinforced Plastics*, *53*(3), 14–21. https://doi.org/10.1016/S0034-3617(09)70078-5.

SukanchanPalit, & MustansarHussain, C. (2018). Handbook of Nanomaterials for Industrial Applications. In C. M. Hussain (Ed.), *Handbook of Nanomaterials for Industrial Applications* (pp. 3–12). Elsevier Ltd. https://doi.org/10.1016/B978-0-12-813351-4.00001-8.

Swain, S. (2013). Synthesis and characterization of graphene based unsaturated polyester resin composites. *Transactions on Electrical and Electronic Materials*, *14*(2), 53–58. https://doi.org/10.4313/TEEM.2013.14.2.53.

Swikker, K. R. J., Kanagasabapathy, H., & Neethi Manickam, I. (2022). Effect of MWCNT on mechanical characterization of glass/carbon hybrid composites. *Proceedings of the*

Institution of Mechanical Engineers, Part E: Journal of Process Mechanical Engineering, *236*(5), 2155–2163. https://doi.org/10.1177/09544089221083464.

Thakur, R. K., & Singh, K. K. (2020). Experimental investigation and optimization of abrasive water jet machining parameter on multi-walled carbon nanotube doped epoxy/carbon laminate. *Measurement: Journal of the International Measurement Confederation*, *164*, 108093. https://doi.org/10.1016/j.measurement.2020.108093.

Tshai, K. H. Y., Khalili, P., Kong, I., Yeoh, C. H., & Tshai, K. H. Y. (2016). Synthesization of graphene and its incorporation into natural fiber reinforced thermosetting nanocomposite. *ARPN Journal of Engineering and Applied Sciences*, *11*(1), 121–127.

Umer, R. (2018). Manufacturing and Mechanical Properties of Graphene Coated Glass Fabric and Epoxy Composites. *Journal of Composites Science*, *2*(2), 17. https://doi.org/10.3390/jcs2020017.

Vaganov, G., Yudin, V., Vuorinen, J., & Molchanov, E. (2016). Influence of multiwalled carbon nanotubes on the processing behavior of epoxy powder compositions and on the mechanical properties of their fiber reinforced composites. *Polymer Composites*, *37*(8), 2377–2383.

Vajaiac, E., Palade, S., Pantazi, A., Stefan, A., Pelin, G., Baran, D., Ban, C., Purica, M., Meltzer, V., Pincu, E., Berbecaru, C., & Dragoman, D. (2015). Mechanical properties of multiwall carbon nanotube-Epoxy composites. *Digest Journal of Nanomaterials and Biostructures*, *10*(2), 359–369.

Vázquez-Moreno, J., Sánchez-Hidalgo, R., Sanz-Horcajo, E., Viña, J., Verdejo, R., & López-Manchado, M. (2019). Preparation and Mechanical Properties of Graphene/Carbon Fiber-Reinforced Hierarchical Polymer Composites. *Journal of Composites Science*, *3*(1), 30 (1-8).

Wang, F., Drzal, L. T., Qin, Y., & Huang, Z. (2016). Enhancement of fracture toughness, mechanical and thermal properties of rubber/epoxy composites by incorporation of graphene nanoplatelets. *Composites Part A: Applied Science and Manufacturing*, *87*, 10–22.

Wang, R., Li, Z., Liu, W., Jiao, W., Hao, L., & Yang, F. (2013). Attapulgite-graphene oxide hybrids as thermal and mechanical reinforcements for epoxy composites. *Composites Science and Technology*, *87*, 29–35.

Watson, G., Starost, K., Bari, P., Faisal, N., Mishra, S., & Njuguna, J. (2017). Tensile and Flexural Properties of Hybrid Graphene Oxide / Epoxy Carbon Fibre Reinforced Composites. *IOP Conference Series: Materials Science and Engineering*, *195*(1). https://doi.org/10.1088/1757-899X/195/1/012009.

Wu, C., Chen, Z., Wang, F., Hu, Y., Wang, E., Rao, Z., & Zhang, X. (2019). Mechanical and drilling properties of graphene oxide modified urea-melamine-phenol formaldehyde composites reinforced by glass fiber. *Composites Part B: Engineering*, *162*, 378–387.

Wu, D., Larsson, R., & Rouhi, M. S. (2019). Modeling and Experimental Validation of the VARTM Process for Thin-Walled Preforms. *Polymers*, *11*(12). https://doi.org/10.3390/polym11122003.

Xiao, C., Tan, Y., Wang, X., Gao, L., Wang, L., & Qi, Z. (2018). Study on interfacial and mechanical improvement of carbon fiber/epoxy composites by depositing multiwalled carbon nanotubes on fibers. *Chemical Physics Letters*, *703*, 8–16. https://doi.org/10.1016/j.cplett.2018.05.012.

Xu, Y., & Hoa, S. Van. (2008). Mechanical properties of carbon fiber reinforced epoxy/clay nanocomposites. *Composites Science and Technology*, *68*(3–4), 854–861.

Yang, S. Y., Lin, W. N., Huang, Y. L., Tien, H. W., Wang, J. Y., Ma, C. C. M., Li, S. M., & Wang, Y. S. (2011). Synergetic effects of graphene platelets and carbon nanotubes on the mechanical and thermal properties of epoxy composites. *Carbon*, *49*(3), 793–803.

Yavari, F., Rafiee, M. A., Rafiee, J., Yu, Z. Z., & Koratkar, N. (2010). Dramatic increase in fatigue life in hierarchical graphene composites. *ACS Applied Materials and Interfaces*, *2*(10), 2738–2743.

Yildirim, F., Aydin, M., & Avci, A. (2021). Improved mechanical performance of three-dimensional woven glass/epoxy spacer composites with carbon nanotubes. *Journal of Reinforced Plastics and Composites*, *40*(13–14), 533–549. https://doi.org/10.1177/0731684421990893.

Yip, M. C., Lin, Y. C., & Wu, C. L. (2011). Effect of multi-walled carbon nanotubes addition on mechanical properties of polymer composites laminate. *Polymers and Polymer Composites*, *19*(2–3), 131–140.

Yu, Z., & Lau, D. (2017). Evaluation on mechanical enhancement and fire resistance of carbon nanotube (CNT) reinforced concrete. *Coupled Systems Mechanics*, *6*(3), 335–349. https://doi.org/10.12989/csm.2017.6.3.335.

Yuan, X., Zhu, B., Cai, X., Qiao, K., Zhao, S., & Zhang, M. (2019). Nanoscale toughening of carbon fiber-reinforced epoxy composites through different surface treatments. *Polymer Engineering and Science*, *59*(3), 625–632. https://doi.org/10.1002/pen.24978.

Zaghloul, M. Y. (2017). Experimental and modeling of mechanical-electrical behavior of polypropylene composites filled with graphite and MWCNT fillers. *Polymer Testing*, *63*, 467–474. https://doi.org/10.1016/j.polymertesting.2017.09.009.

Zakaria, M. R., Abdul Kudus, M. H., Md. Akil, H., & Mohd Thirmizir, M. Z. (2017a). Comparative study of graphene nanoparticle and multiwall carbon nanotube filled epoxy nanocomposites based on mechanical, thermal and dielectric properties. *Composites Part B: Engineering*, *119*, 57–66.

Zakaria, M. R., Abdul Kudus, M. H., Md. Akil, H., & Mohd Thirmizir, M. Z. (2017b). Comparative study of graphene nanoparticle and multiwall carbon nanotube filled epoxy nanocomposites based on mechanical, thermal and dielectric properties. *Composites Part B: Engineering*, *119*, 57–66.

Zhang, L., Jiao, H., Jiu, H., Chang, J., Zhang, S., & Zhao, Y. (2016). Thermal, mechanical and electrical properties of polyurethane/(3-aminopropyl) triethoxysilane functionalized graphene/epoxy resin interpenetrating shape memory polymer composites. *Composites Part A: Applied Science and Manufacturing*, *90*, 286–295.

Zhang, R. L., Gao, B., Du, W. T., Zhang, J., Cui, H. Z., Liu, L., Ma, Q. H., Wang, C. G., & Li, F. H. (2016). Enhanced mechanical properties of multiscale carbon fiber/epoxy composites by fiber surface treatment with graphene oxide/polyhedral oligomeric silsesquioxane. *Composites Part A: Applied Science and Manufacturing*, *84*, 455–463. https://doi.org/10.1016/j.compositesa.2016.02.021.

Zhang, X., Chen, H., & Zhang, H. (2007). Layer-by-layer assembly: From conventional to unconventional methods. *Chemical Communications*, *14*, 1395–1405. https://doi.org/10.1039/b615590a.

Chapter 3

Evaluation of Machining Characteristics of Laminated Polymer Nanocomposites: Conventional Machining

3.1. Problem Formulation

Polymer composites (PCs) have been one of the most practical and cost-effective options for replacing heavyweight materials with improved mechanical and electrical properties since the 21st era. The qualities of polymer-based composites include higher strength at a lower weight, corrosion and moisture resistance, and wear resistance. To meet the different needs of consumers, the manufacturing industry emphasizes producing such materials. Reinforcements (fiber) are commonly employed in the polymer industry to improve the characteristics of composite materials (Caggiano et al., 2018; Sanjay & Yogesha, 2018; Srivastava et al., 2020). The potential for additional development exists in multifunctionality, as reinforcing agents are employed to improve the desired qualities. Fiber-reinforced polymer (FRP) production can benefit the most from introducing nanoparticles. Various fillers and doping nanomaterials are incorporated into these materials to modify them. Engineering applications choose nanoparticles because they enhance the polymer matrix phase's synergistic effect. In this series, epoxy matrix materials frequently incorporate nano-reinforcing materials, including carbon nanotubes (CNT), carbon nano onions (CNO), carbon nanorods (CNR), graphene, and others (Papageorgiou et al., 2020). For end-use, innovative materials must be machined and machinable. Consequently, machining research is the primary focus of trade and academia. It requires unique machining approaches compared to traditional material properties like metal alloys, non-metallics, etc. The ability to regulate process parameters significantly impacts the surface characteristics, aesthetic requirements, and production rate of components or samples that need to be machined. Drilling, milling, and grinding are the basic machining operations used in the manufacturing industries to produce the numerous parts required for the finished product's dimensional assembly (El-Sonbaty et al., 2004; P. Li et al., 2019; Zarif Karimi et al., 2017). The drilling process is required to produce holes, rivets, and fastening assembly of a high quality and degree of precision. Machined holes with poor surface quality can hamper assembly performance. Furthermore, after a certain period, it may diminish the product's strength, efficiency, and durability (Brinksmeier, 1990; Saoudi et al., 2016; Sorrentino et al., 2017; Zitoune et al., 2016). CNC drilling is a standard production process often used instead of conventional machining methods for various materials (Caggiano, 2018). The drilling process is commonly used to make a hole in the material with exact specifications for different industrial purposes. Acknowledging the several efforts made by eminent researchers, cutting the holes accounts for 40% of assembly and production time in the manufacturing industries (Liu et al., 2012; Saoudi et al., 2016; Zitoune et al., 2016). Machining CFRP composites differs significantly from processing metal because of plastic deformation and inhomogeneous structure (Ha et al., 2017; Kaybal et al., 2019; Kurniawan et al., 2017). Polymer is machined using conventional methods, such as CNC drilling, to generate a precise hole with the needed quality. Researchers have made incredible strides in polymer processing in the past two decades. It is evident from the available data that there is a lack of investigation on the machining characteristics of nanocomposites. This chapter presents the findings of the drilling and milling tests conducted on modified epoxy-laminated nanocomposites and a study of their machineability. Adjusting process parameters has been investigated to reduce damage and defect rates. Here, it is explored that nanofillers affect machining efficiency and various methods to improve machinability. The proposed nanocomposites' capacity to create slots and holes through milling and drilling operations has been examined.

Drilling and Milling are the primary machining processes for creating holes and slots. The production of polymers is a fundamental machining process, and it may be studied using the appropriate set of parameters (set of tools) (Tsao, 2008). Due to tool materials and geometry, little heat is required during machining to achieve surface integrity, high tool life, and high cutting rates. Over this, drilling and milling PCs present a considerable machining challenge. While drilling polymeric materials, the quality of the hole was affected by the design parameter of the tool and other process variables (Dong Wang et al., 2019a). Many experts have studied the machinability of carbon/PCs and provided numerous concepts and tools to improve the machining process. While traditional drilling and milling are the most popular machining used on FRP composites, different types are found (Komanduri, 1997). The primary research focuses on the polymer-based structural machining investigation of tool geometry and materials, tool types, parameters, etc. Figure 3.1 demonstrates the schematic drilling operation setup. The cutting force (Cf) is crucial in defining the material removal rate (MRR) during machining processes. It directly impacts factors like product finish, durability, process energy, cost of process, and tool life (Bhat et al., 2019; Mohan et al., 2005). There is currently a preliminary phase of information on the machining process of graphene oxide/carbon fiber (GO/CF) polymer nanocomposites. Due to plastic deformation and its non-homogeneity, machining it is challenging. Furthermore, due to the complex anisotropic nature of CFRP, Cf, and tool wear rate (TWR) is higher than in other materials (Miyake et al., 2016; Rao et al., 2007). Numerous studies have demonstrated that the feed rate (F) is the most influential component while machining composite materials (Abrão et al., 2008; Tsao, 2008). The resistance force increases due to the increased rubbing of the drill tool interfaces on the inner surface of the hole, caused by the high impact of the cutting tool edges against the abrasive fibres. The increased speed is the primary cause (Janakiraman et al., 2020; Park et al., 2018). It is also challenging to make a precise and accurate end product in polymer nanocomposites without considering the influence of Cf.

3.2. Experimentation Details and Investigation of Drilling Responses

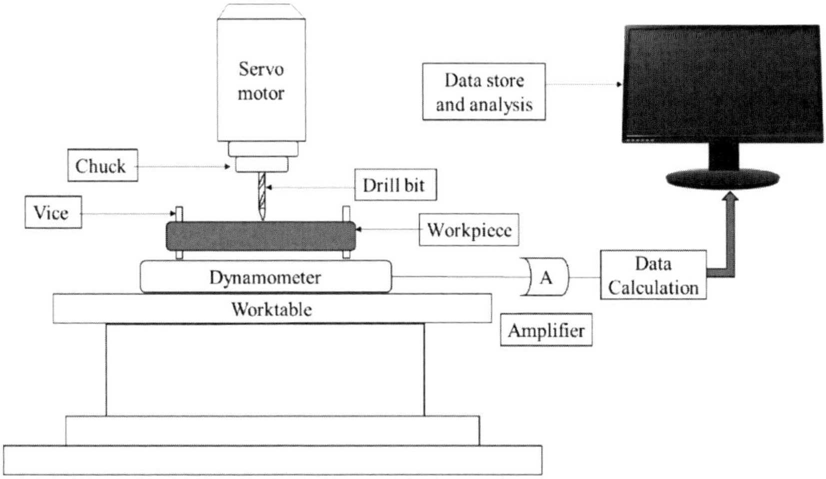

Figure 3.1. Schematic diagram of drilling setup (J. Kumar, Verma, & Mondal, 2021).

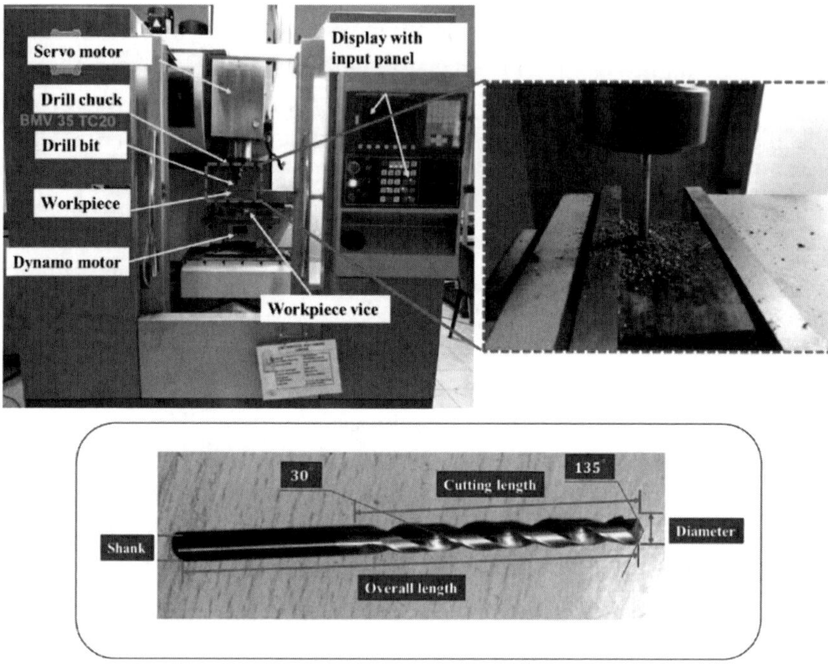

Figure 3.2. Vertical CNC machine and drill tool (J. Kumar et al., 2020).

Table 3.1. CNC Machine

Model number	Feed	Spindle speed	Transvers (X, Y and Z axis)
BMV35 TC20	1-10000 mm/min	10000 rpm	450, 350, and 350 mm

Table 3.2. Dynamometer

Model number	Measuring range	Sensitivity
PML-300	4 KN	-26 pc/N

This section provides brief information about the machine setup, tool, and condition used in this work for machining and machinability (Drilling process) performance. For polymer nanocomposite machining, the drilling response measurement method plays a prominent role and is generally considered standard. TiAlN (SiC coated) drill bits (5 mm) were used in experiments on a CNC vertical drilling machine (Model No. BMV35 TC20) equipped with a tool dynamometer (Model No. MLB-PML-300). Using the proper clamping arrangement, the sample was clamped securely in the CNC machining center.

The experimental drilling, shown in Figure 3.2, included a 5 mm diameter TiAlN-coated SiC drill bit and an appropriate clamping system for holding the workpiece to the center. Interfaces between coating layers impart increased hardness and durability. The drill tool used to make the hole in the composite had a point angle of 135 degrees and a helix angle of 30 degrees. The machining response was thrust force (Th), torque (T), surface roughness (Ra), and delamination (Fd), and the process parameters were spindle speed (S), F, and GO wt.%. Tables 3.1 and 3.2 detail the most fundamental features of the CNC machine and dynamometer, respectively.

3.2.1. Surface Roughness of Machined Samples

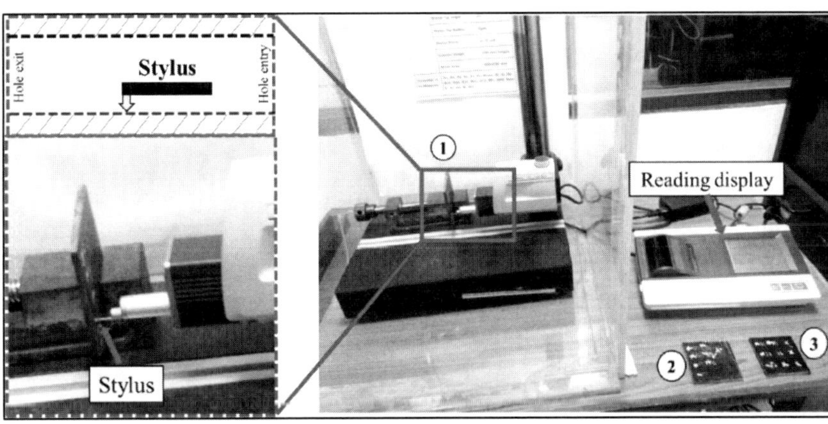

Figure 3.3. Ra measurement setup (J. Kumar & Verma, 2021a).

Table 3.3. Ra tester

Model number	Range	Resolution	Measuring force	Measuring speed
Mitutoyo SJ-400	8 μm	0.000125μm	0.75mN	1.0 mm/sec

The surface morphologies of machined holes were evaluated over a range of process parameters using the roughness tester (Mitutoyo SJ-400), as displayed in Figure 3.3. The two most common indices (Ra-average roughness and Rz-10-point scale roughness) have been applied to each machined hole for surface finishing evaluation. The roughness of the surface was evaluated by employing a rectangular probe with a 2 m tip radius and a length of 0.8 mm. A 5 mm sampling length has been set for each sample hole surface of the machined layers that moved toward drilling. Three measurements of the hole's Ra were collected, and the average was used to calculate the surface indices. Fiber fracture and burr formation contribute to unevenness of the machined surface. The stylus went around the entire hole so that the perpendicular motion caused by the surface variations could be measured. The vertical motion is translated into a more significant electric signal. After filtering the signal, the Ra was calculated, and the resulting value was shown on the roughness tester digital screen. Table 3.3 depicts the features of Ra.

Figure 3.4 depicts the average Ra of the laminated polymer nanocomposite (LPNC) versus the experimental constraints. For twist drill bits in the cutting set, the average Ra value of the composite was 1.65 micrometers. Ra indices evaluated after the composites cutting phase were lower for LPNC than for pure laminated drilling. This occurrence demonstrated that drilling could improve the quality of machined holes. Drilling with less friction is possible due to the TiAlN coating's unique anti-friction chip interaction. In addition, all experiments showed that the feed had a notable effect on the roughness of the drilled hole, with higher feed values producing noticeably rough properties. In contrast, at the most significant F (240 mm/min), the typical maximum Ra value recorded in machined holes is 2.0325 μm. There was some variation in surface quality, although it had to be minimally affected by the rotation of the tool speed. The findings of the comparisons also demonstrated that a drilling LPNC cutting sequence characterized by a lower F and a higher S might be employed to achieve a higher-quality surface finish. The Ra of machined holes is most affected by the F, as was also suggested by Ogawa et al., (Ogawa et al., 1997). The results of the machined holes are shown in Figure 3.4, along with the same observation made concerning the colored surface at varying wt.% of Nanomaterials. GO at 1 wt. % produced the minimum roughness (Ra=1.65 m). As shown in Figure 3.4, the surface profile of high feeds containing a high % of Nanomaterials is rough due to the creation of large voids and fibre pull-outs. However, the likelihood of fiber/matrix debonding, fiber pull-outs, and cavities increases significantly during processing under these conditions. The uneven fibre and matrix fractures may result from the increased pressure imposed by more significant F. As previously noted, the quality of the surface can also be considerably enhanced by machining at lower F and higher S (D. Kumar & Gururaja, 2020). The existence of cavities, de-cohesion of the fiber matrix, and fiber/matrix dust can lead to poor surface quality. While the F was reduced and the S increased, the surface quality improved dramatically (Gupta & Kumar, 2015; Xu et al., 2019).

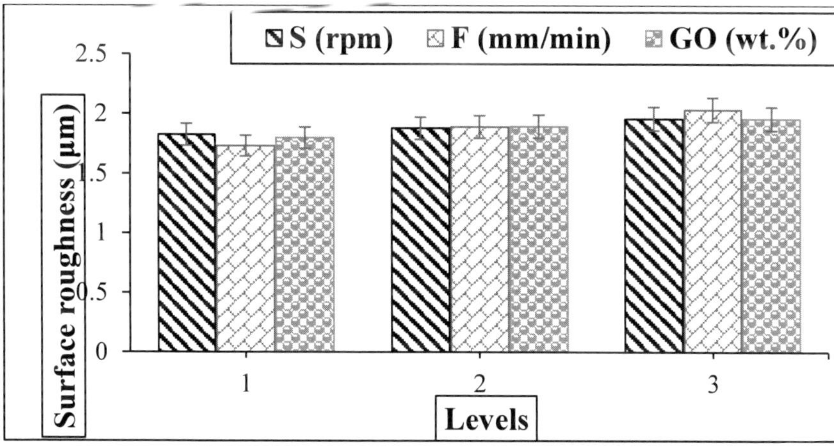

Figure 3.4. Ra Vs. Drilling factors analysis.

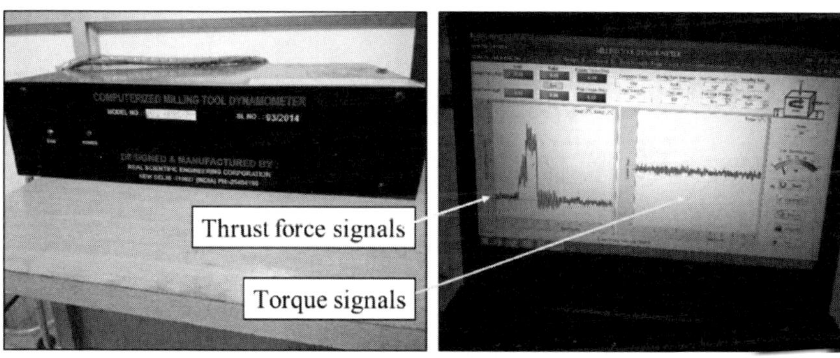

Figure 3.5. Measurement of Th and T (J. Kumar & Verma, 2021c).

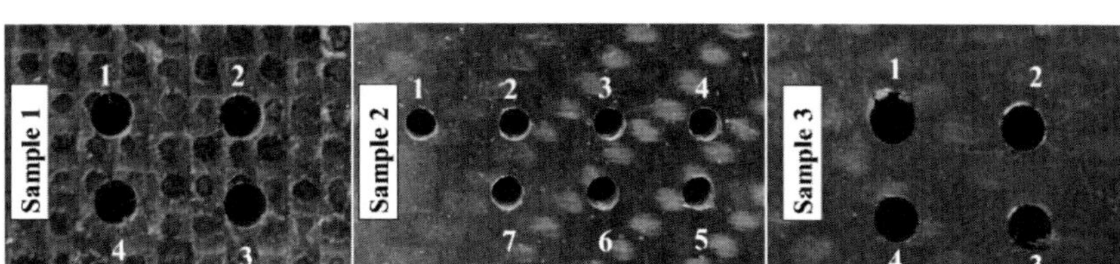

Figure 3.6. Drilled sample.

3.2.2. Thrust Force and Torque

An MLB-PML-300 tool dynamometer was attached to a CNC vertical drilling machine to evaluate the measured data (Figure 3.5). A computerized dynamometer was employed to capture online data for T and Th. The online monitoring of the GO/CF nanocomposite drilling process was investigated to better understand the workpiece's current state and tool interactions. Signals generated from the cutting time were utilized to analyze the drilling force. The Th and T signals were extracted from the drilling force data. It is proposed that different machining conditions be compared regarding the force signals they produce. The microscope image of the drilling hole is displayed in Figure 3.6.

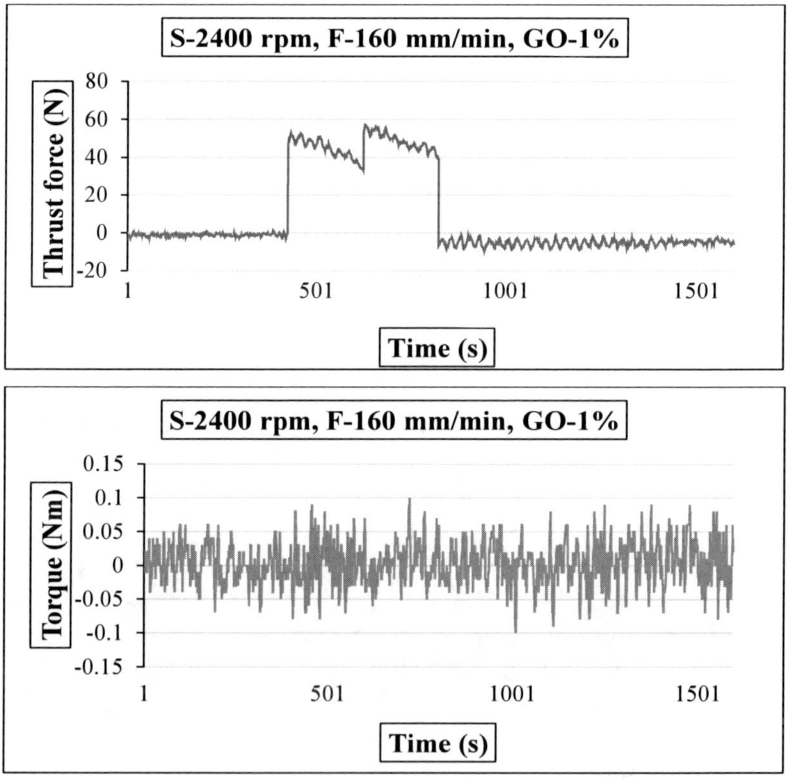

(a)

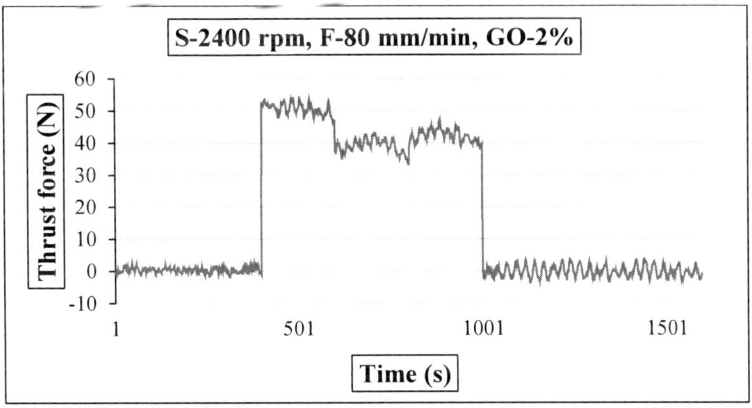

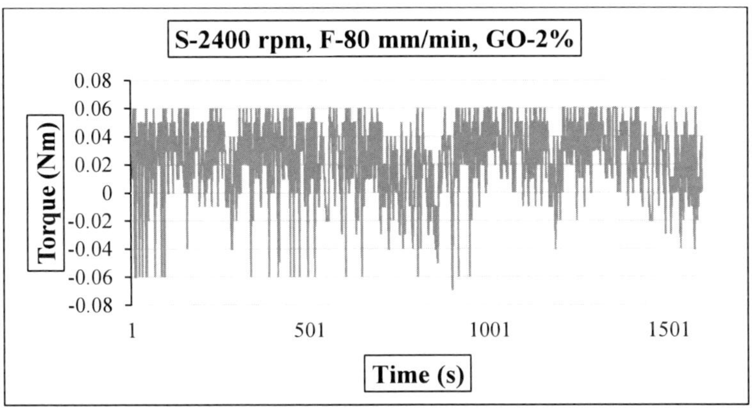

(b)

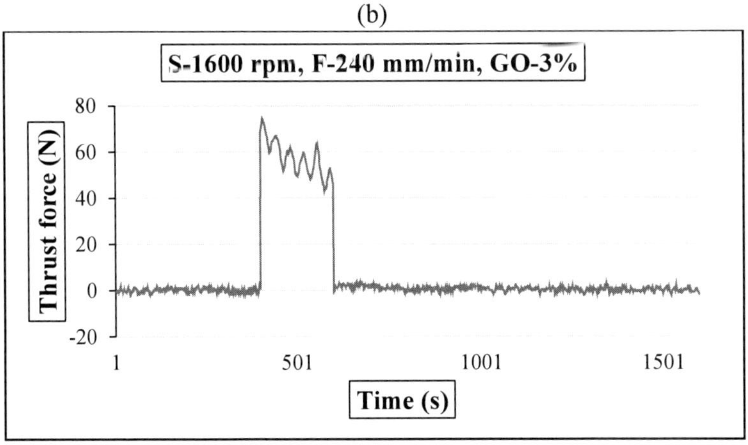

Figure 3.7. (Continued).

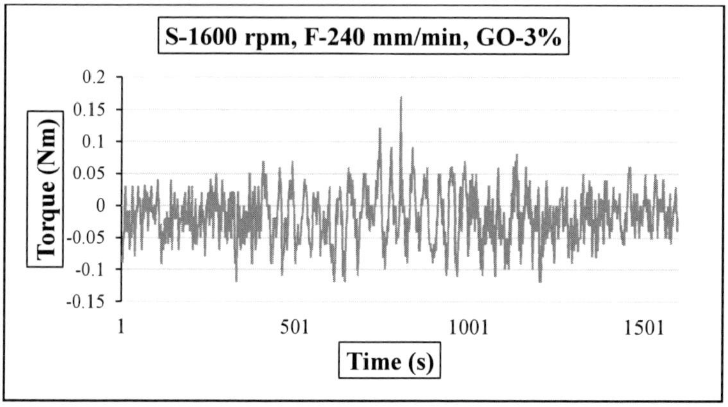

(c)

Figure 3.7. Examination of T and Th signals (a) S-2400 rpm, F-160 mm/min, GO-1%, (b) S-2400 rpm, F-80 mm/min, GO-2%, (c) S-1600 rpm, F-240 mm/min, GO-3% (J. Kumar & Verma, 2021c).

The Th and T play a crucial role in ensuring the PC's processability and producing a high-quality machined product. T_h and T were assessed in real-time while drilling the LPNC using a dynamometer-CNC attachment. Figures 4.7 (a-c) show the progression of T_h and T during the machining of CNMs containing 1%, 2%, and 3% doped nanocomposites, respectively. The drilling process consists of three steps to generate a through hole. The workpiece touches and begins to penetrate during the first step, resulting in a constant T_h value (a-c).

The second phase's highest peak point on the workpiece experienced a more excellent value of Th when the whole cutting edge contacted it. In the third stage, the drill exits or splits the workpiece, generating the constant Th again. In the third phase, the T_h values are significantly reduced. The machining process is likely to finish there since there is no longer any significant contact between the workpiece and the tool (D. Kumar & Singh, 2019). The investigation showed that drilling caused Th to range from 53.94 N at F-80 mm/min to 100.81 N at F-240 mm/min, with T ranging from 0.06 Nm to 0.17 Nm. The influence of Th and T was found to be the primary source of drilling-induced Fd in most of the investigations (Krishnamoorthy et al., 2012). Higher Th and T produced during drilling typically cause machine tool failure through composites (Alagumurthi et al., 2017). S and F, among other process constraints, can be tuned to achieve the desired results (Shunmugesh et al., 2017). Figures 4.7 (a-c) show that the F significantly impacts the effects of the Th and T forces. F, higher than those required for cutting, will cause the drilling tool to penetrate rather than cut through the layer. With higher F, Th and T increases. Several research studies indicate that PCs' machining performance is most affected by F (Abrão et al., 2008; Tsao, 2008). As a result of the abrasive fibers having a substantial effect on the cutting edges, the resistance force and rubbing of the drill tool against the inside of the hole increases (Janakiraman et al., 2020; Park et al., 2018).

Using a twist drill bit to drill LPNC, Figures 3.8 and 3.9 demonstrate the Th and T impact as a function of cutting parameters. It was observed that the F significantly affected Th during the drilling test. The higher shear zone becomes more intense as feed levels rise. At an F of 80 mm/min, the average Th was 58.44 N, while at a 240 mm/min rate, the average Th was 78.67 N. However, the Th power varies slightly depending on the cutting speed of the twist drill used for LPNC drilling. However, the force required to cut through this material reduces at higher speeds (Wu et al., 2019). Matrix softening is shown as a result of the high cutting speed increasing the temperature of the cutting drills. T falls progressively with decreasing F, producing an effect like that of the Th under optimum operation conditions. There was also a linear relationship between the F and the Th and T, demonstrating the importance of F in the drilling operation (D. Kumar & Singh, 2015). For a reduced Th and T result, optimal temperatures with an increase in heat generation due to a low thermal coefficient and a low plastic transition temperature were required (Wei et al., 2016).

3.2.3. Delamination at Inlet and Outlet

As shown in Figure 3.10, the Fd value was determined using an optical microscope with a magnification range of 0.7x-4.5x and a precision of 0.0001mm. In this study, Fd was measured using the available, open-source MSU-3D Pro software for image processing. All damages are assessed by locating distinct spots across the entry and exit sides of the drilled hole.

Equation (5.1) is typically employed to determine the Fd, particularly the ratio of the nominal hole diameter to the maximum diameter of the damaged region. The image processing method used to determine the Fd effect is depicted in Figure 3.11.

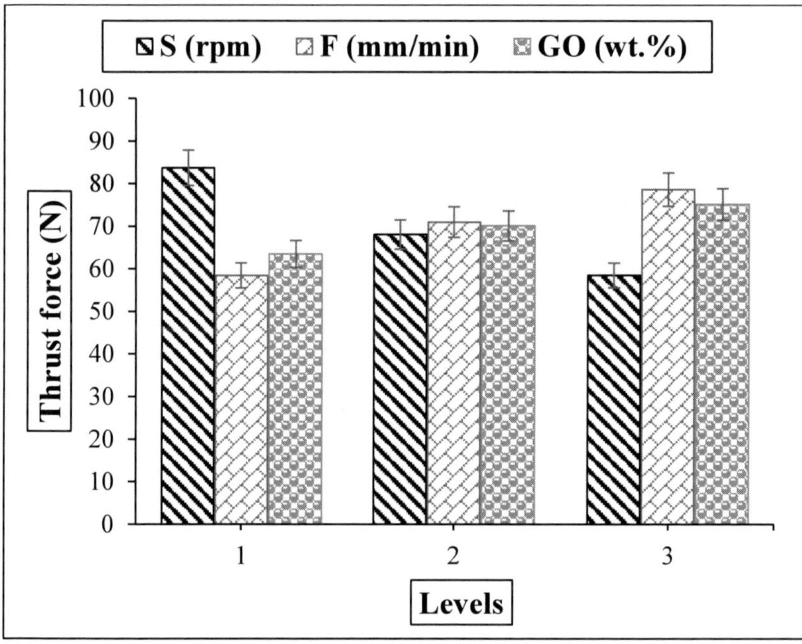

Figure 3.8. Th Vs. Drilling factors analysis.

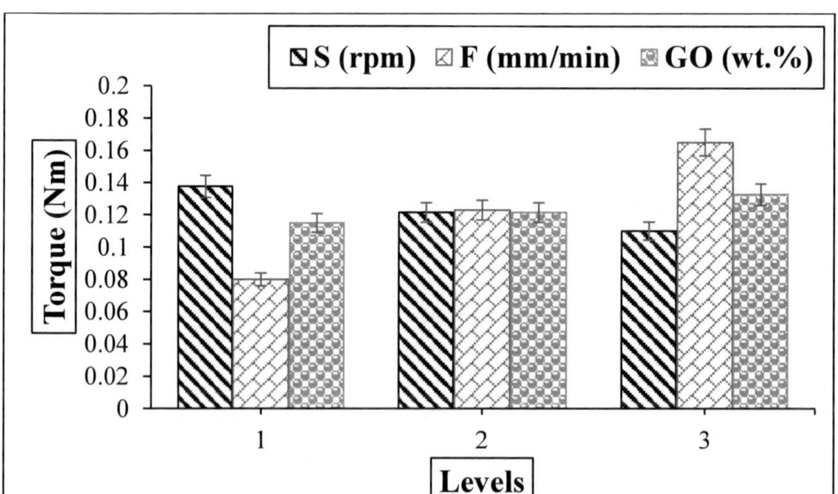

Figure 3.9. T force Vs. Drilling factors analysis.

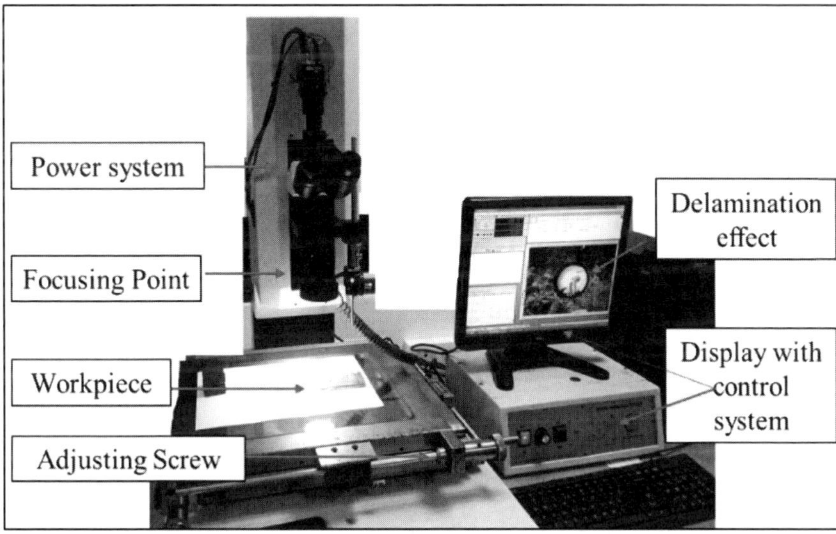

Figure 3.10. Vision measurement system (J. Kumar et al., 2020).

$$F_d = \frac{D_{max}}{D} \tag{5.1}$$

Here, F_d = Delamination factor, D_{max} = Maximum damages diameter of drill hole, and D = Nominal diameter of the drill hole.

The inlet and outlet hole microscopic images are displayed in Figure 3.12 and Figure 3.13.

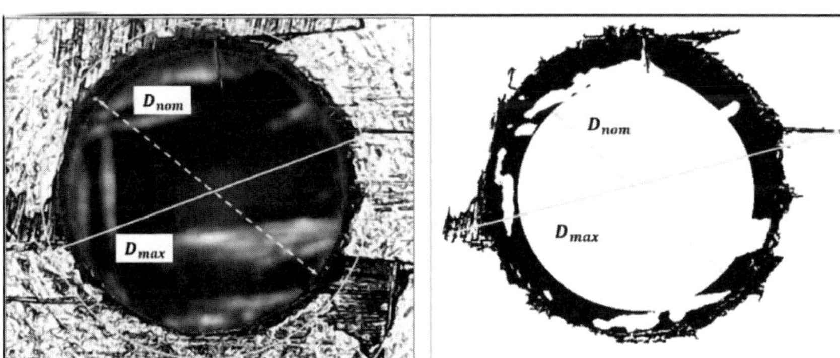

Figure 3.11. Fd effect (J. Kumar & Verma, 2024).

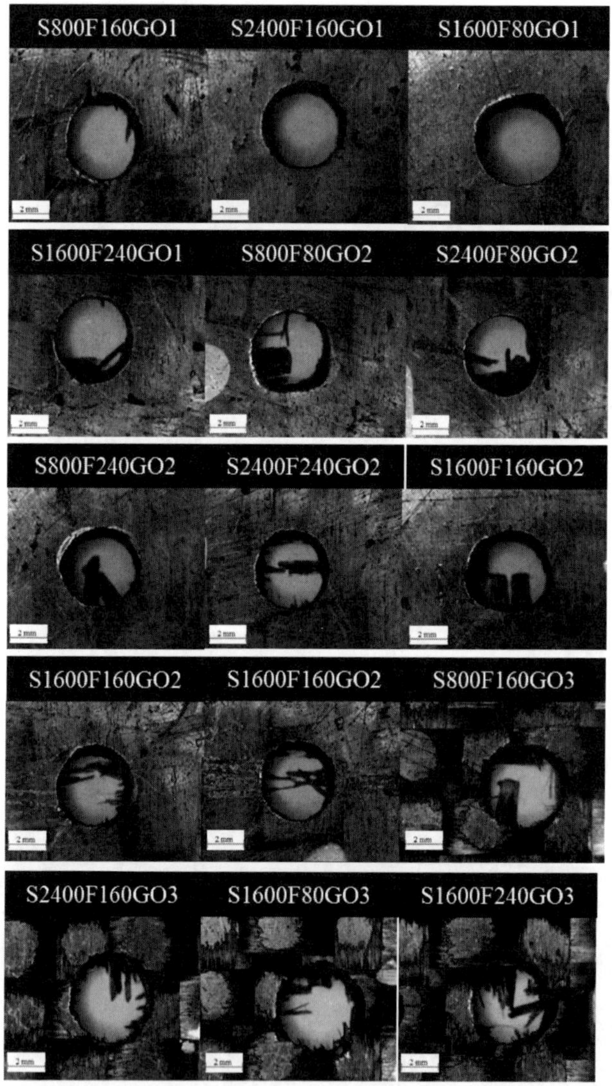

Figure 3.12. Fd effect at inlet using microscopic images (J. Kumar et al., 2020).

Drilling-induced Fd is a significant challenge for manufacturers of CFRP nanocomposites. An image processing method and a microscope determined the size of the hole's damage. As previously discussed, Table 3.4 reports on the selection of process parameters and their level for drilling experimentation. The drilling-induced Fd at the inlet (IN) and outlet (OUT) is also reported.

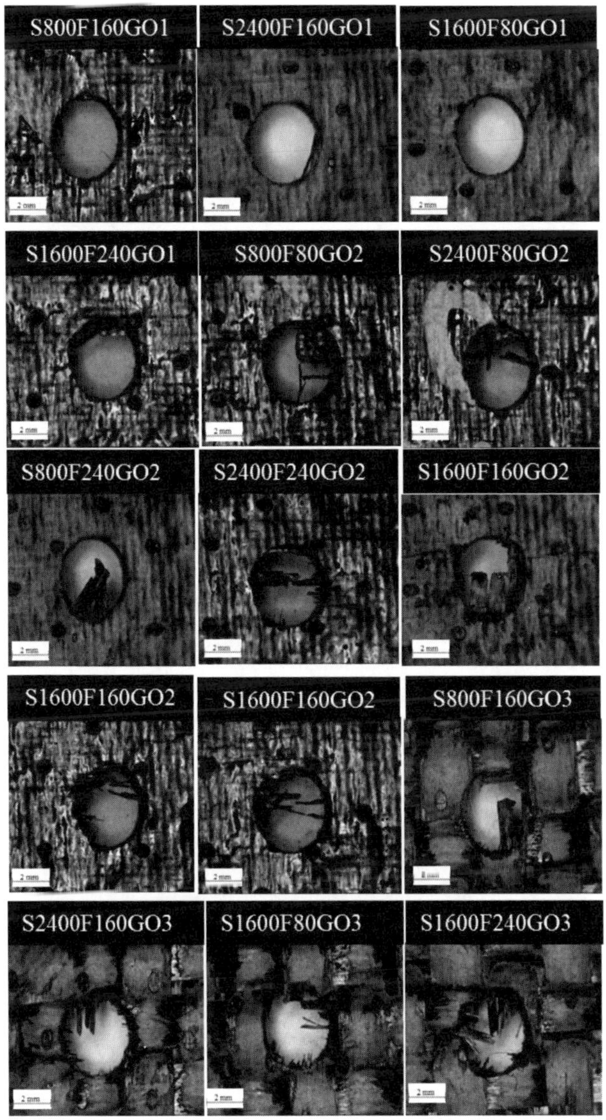

Figure 3.13. Fd effect at exit using microscopic images (J. Kumar et al., 2020).

Drilling of polymer materials revealed deterioration on the entrance sides. Damage mechanisms, such as peeling in drilled CFRP laminated nanocomposite materials are readily apparent. During drilling, Th is the primary component that causes damage (Hocheng & Tsao, 2003). Fd (IN) occurs when the cutting tool of a drill contacts the laminate. Drill flutes then pull off the upper lamina, which was nibbled away initially. The uncut layers can be elastically bent because the compressive strength of the drill bit causes the thickness of the uncut laminates to decrease along the bottom edge (Dong Wang et al., 2019b). Figure 3.14 and Figure 3.15 show the process parameters influencing the Fd (IN). As the percentage of GO in CFRP nanocomposites increases, the Fd decreases on the intake side. Increases in S also reduce the amount of Fd at the entrance. The lowest average Fd (1.025) was found at 1 wt. % GO supplement with CFRP. Strength at the reinforcing epoxy interface is increased when GO is included in the matrix. Using it during machining offers lubrication at the point where the tool's and the chip's cutting edge meet and lowers friction (Fu et al., 2019; He et al., 2018). It demonstrates that a slight increase in the exit side delamination factor can be attributed to a greater F during machining (Boughdiri et al., 2021; Gaitonde et al., 2008; Isbilir & Ghassemieh, 2012; D. Kumar & Singh, 2019; Meinhard et al., 2021; Sridharan et al., 2016). Increased F increases Th which can bend uncut materials elastically and raise bending stresses on composites. Laminar splitting is possible since it will decrease interlaminar shear strength (ILSS) (Adak et al., 2018). As the drilling bit goes downward, the inter-laminar bond failure causes cracks to form and spread around the hole. The modulus of elasticity strength and ILSS can be increased by adding CNMs to epoxy (L. Li et al., 2019; Qin et al., 2015; Vijayan & Rajmohan, 2019).

Table 3.4. Observation of drilling induced Fd (IN/OUT) (J. Kumar and Verma, 2021c)

Exp. No.	Process parameters			Fd	
	S (rpm)	F (mm/min)	GO (Wt.%)	IN	OUT
1	800	80	2	1.031	1.043
2	2400	80	2	1.012	1.022
3	800	240	2	1.095	1.111
4	2400	240	2	1.040	1.061
5	800	160	1	1.040	1.066
6	2400	160	1	1.020	1.040
7	800	160	3	1.060	1.091
8	2400	160	3	1.030	1.053
9	1600	80	1	1.012	1.026
10	1600	240	1	1.051	1.072
11	1600	80	3	1.030	1.043
12	1600	240	3	1.071	1.118
13	1600	160	2	1.060	1.081
14	1600	160	2	1.050	1.064
15	1600	160	2	1.036	1.064

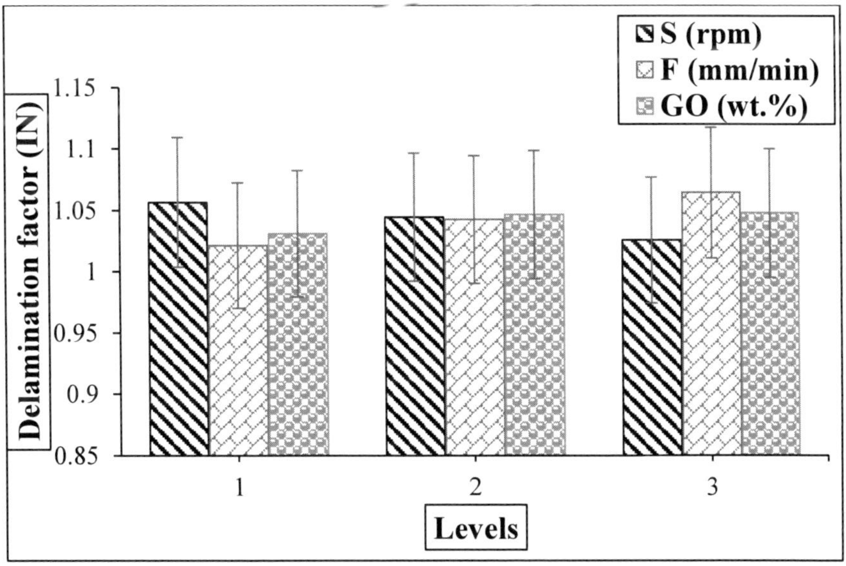

Figure 3.14. Fd (IN) Vs. Drilling factors analysis.

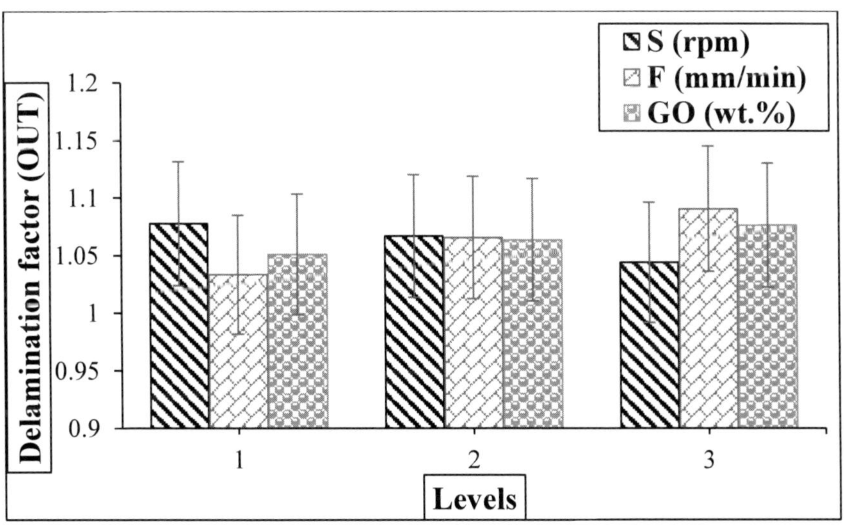

Figure 3.15. Fd (OUT) Vs. Drilling factors analysis.

3.2.4. Circularity Error

It measures the distance around two concentric circles that fit the data to find the circularity error (Ce) for a set of measurement points. Identifying an assessment function that is superimposed on the actual value under consideration is advantageous to obtain the smallest feasible value of the most significant distance from it (Samuel & Shunmugam, 2003). Although machine tools often have round shapes, accurate measurement of circularity inaccuracy is crucial (internal and external). It can be calculated using two general methods: the intrinsic datum technique and the extrinsic datum technique. The most often used instruments are bench center measurements, V block measurements, and diametrical assessments. Diametric measurement instruments like micrometers, air gauges, and bore gauges may measure element diameter in diverse cross-sections. The V-block technique rotates the component in a V block and employs a dial indicator to check for circularity errors. The component is rotated between two precise centers using the bench center method, and the inaccuracy is calculated using a dial indicator. Four common measurement methods fall under the extrinsic methodology category: minimum circumscribed (MC), maximum inscribed (MI), least-square (LS), and minimum radial separation (MRS) circle (Rajagopal & Anand, 1999; M. Wang et al., 1999).

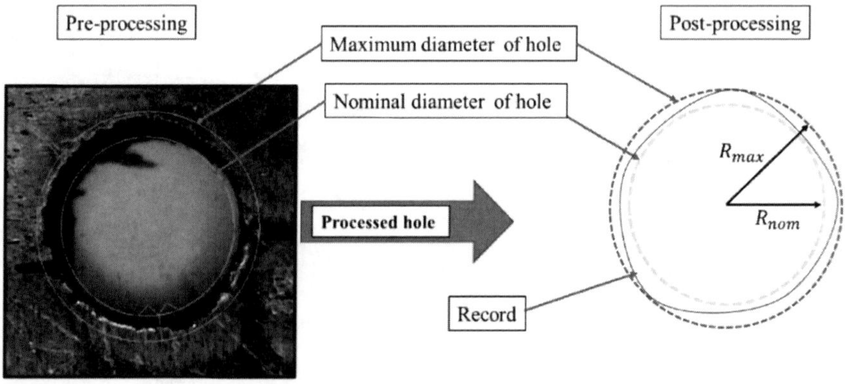

Figure 3.16. Assessment of circularity error (C_e) (J. Kumar & Kumar Verma, 2021).

This study computes vision measuring machine circularity error using the advanced radical technique. Figure 3.16 shows the drilling hole circumference circularity error statistics. For each drill hole, the circularity errors in the machined holes were measured using an exact vision measuring machine setup. Table 3.5 provides a summary of the experiments that correspond to the circularity response values.

Table 3.5. Observation of circularity error (C_e) (J. Kumar and Kumar Verma, 2021)

Exp. No.	Process parameters			Response
	S (rpm)	F (mm/min)	GO (Wt.%)	C_e
1	800	80	2	1.031
2	2400	80	2	1.012
3	800	240	2	1.095
4	2400	240	2	1.040
5	800	160	1	1.040
6	2400	160	1	1.020
7	800	160	3	1.060
8	2400	160	3	1.030
9	1600	80	1	1.012
10	1600	240	1	1.051
11	1600	80	3	1.030
12	1600	240	3	1.071
13	1600	160	2	1.060
14	1600	160	2	1.050
15	1600	160	2	1.036

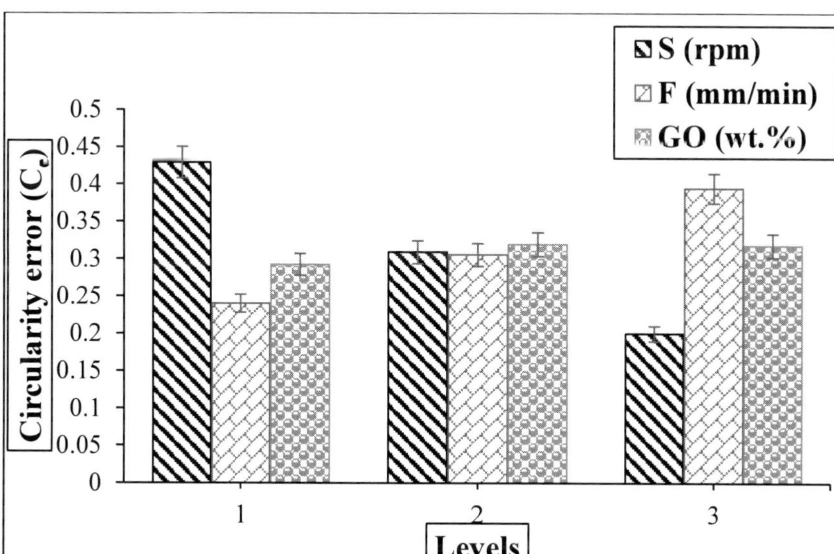

Figure 3.17. Circularity error Vs. Drilling factors analysis.

Changing the input variables from one level to another makes it possible to assess the impact of various constraints on the circularity error (Ce). The average circularity error value is generated to determine the effects of the parameters under consideration. The results demonstrate that the greater S-2400 rpm could produce the 0.133 mm circularity error. Increasing S, decreasing feed, and decreasing carbon nanomaterial content yields the lowest Ce value. It has shown how the various responses to the process parameters influence the values of Ce. In addition, the average is used to calculate the response's deviation from the normal. The raw data response table results were verified by the graph showing the mean response values. Typical Ce response data is shown on the response graph in Figure 3.17.

3.3. Experimentation Details and Investigation of Milling Responses

Figure 3.18. Machine setup for milling process (J. Kumar, Verma, Mondal, et al., 2021).

Milling is widely employed in manufacturing due to its many benefits, including high product quality, customer focus, low production costs, and simple implementation (Azmi et al., 2013). Maximizing MRR, minimizing Ra, and minimizing Cf are the three key output concerns that determine milling performance. An arrangement of a vertical CNC milling machine (Model No. BMV35 TC20) with a Titanium-Aluminum Nitride (TiAlN) milling tool (diameter, 5 mm) was used for the tests (Figure 3.18) at three different levels of machine restrictions, including cutting speed (Cv), F, depth of cut (D), and content % of carbon nanoparticle (Figure 3.19).

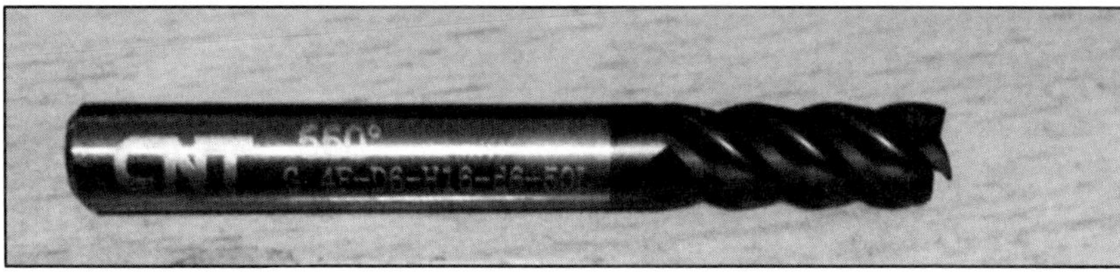

Figure 3.19. Milling cutter (J. Kumar & Verma, 2023).

The experiments were carried out at three distinct levels (Table 3.6) of process limits identified to Vc, F, D, and GO wt.%. The sequence of experiments and associated observational data are listed in Table 3.7.

Table 3.6. Process constraints in milling process (J. Kumar, Verma, Mondal, et al., 2021)

Constraints	Nomenclature	Level		
		1	2	3
Cutting speed (m/min)	V_c	12.56	25.12	37.68
Feed rate (mm/min)	F	80	160	240
Depth of cut (mm)	D	0.5	1	1.5
Graphene oxide (wt.%)	GO	1	2	3

Here, 12.56, 25.12, 37.68 V_c (m/min) is 800, 1600, and 2400 rpm, respectively.

Table 3.7. Observation of milling response (MRR, Cf and R_a) (J. Kumar, Verma, Mondal, et al., 2021)

Exp. No.	Process parameter				Machining response		
	V_c (m/min)	F (mm/min)	D (mm)	GO (wt.%)	MRR (mm^3/min)	*Cf* (N)	R_a (μm)
1	12.56	80	0.5	1	3.793	15.78	1.12
2	12.56	160	1	2	11.78	42.579	1.487
3	12.56	240	1.5	3	11.181	60.075	1.397
4	25.12	80	1	3	10.782	17.757	1.313
5	25.12	160	1.5	1	12.379	16.78	1.483
6	25.12	240	0.5	2	15.973	5.602	0.783
7	37.68	80	1.5	2	9.184	53.675	1.54
8	37.68	160	0.5	3	4.592	9.534	1.02
9	37.68	240	1	1	7.587	4.735	1.93

3.3.1. Material Removal Rate (MRR)

MRR measures the material removed during machining in a specified time. A computerised weighing balance was used to determine the beginning and final weights before and after the experiment.

The materials removal rate (MRR) is calculated by using the following expression (Ponappa et al., 2019):

$$MRR = \frac{w_i - w_f}{\rho \times t_m} \tag{4.1}$$

where:

w_i, w_f = Initial and Final weight of the machined sample, ρ = Density of workpiece (1.31 g/cm^3), and t_m=Machining time.

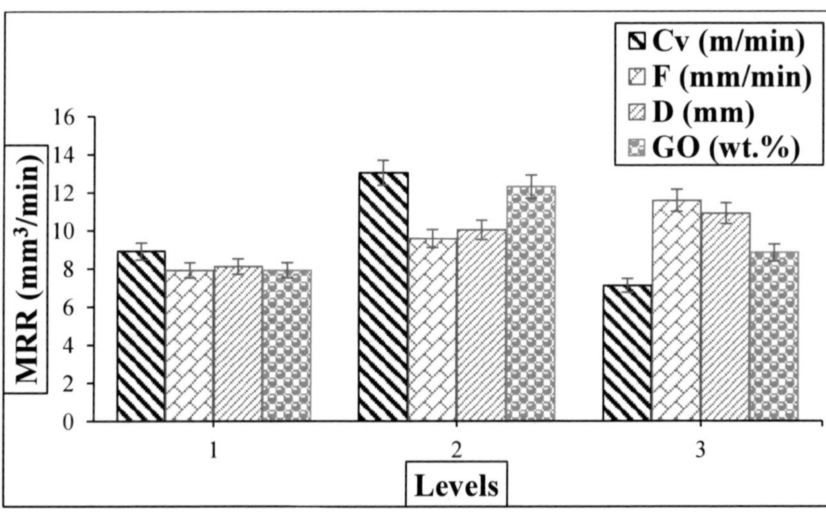

Figure 3.20. MRR Vs. Milling factors analysis (J. Kumar et al., 2022).

In contrast to conventional milling, the cutting speed rose considerably at a regulated MRR. At 25.12 m/min, it revealed a higher result for the MRR. Though its higher level would be more resilient, a greater feed value is recommended for MRR. Compared to the conventional materials, the prevailing MRR dropped at a lower cutting depth. The MRR increased as the cutting depth changed at a greater value, with the F increasing. Successful MRR could be accomplished if the cutting is wholly removed from the fiber, and less force is placed on the tool (Palanikumar et al., 2006). It consequently leads to reduced specific cutting pressure; hence, greater cutting depth is preferred to satisfy the maximum of performance characteristics (Palanikumar et al., 2006). Depending on the characteristics of the Cf, they may indicate changes due to irregularities in cutting speed. LPNC exhibits more ductility in representing the Cf attributes than traditional materials at higher F due to increased temperature and better molecular mobility. TWR is affected by the inadequate heat transfer from the machined sample. Figure 3.20 shows that the MRR was increased dramatically during the machining process with increased D-1.5 mm, F-240 mm/min, Cv-25.12 m/min, and GO-2wt.%. It is found that the MRR significantly increases with the depth of the cut and F at 1.0 mm and 240 mm/min. Also, the medium cutting speed and wt.% of GO significantly improved the MRR.

3.3.2. Cutting Force (Cf)

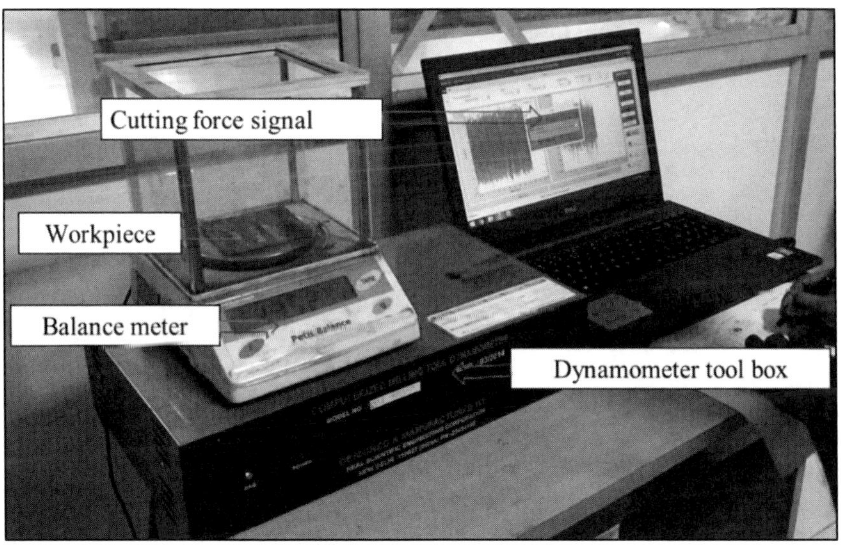

Figure 3.21. Setup for measuring Cf and using a digital balance meter (J. Kumar, Verma, Mondal, et al., 2021).

The computerized dynamometer has been used to obtain data for Th, radial force, and T as displayed in Figure 3.21. Through the analog-to-digital converter, the received signal has been sent to the data acquisition system (Ray et al., 2019). The value of the Cf was calculated by using the expression (Jenarthanan & Jeyapaul, 2018; Pang et al., 2014).

$$F_c = \sqrt{(F_x^2 + F_y^2 + F_z^2)} \tag{4.2}$$

where:

F_x =Axial force, F_y= Radial force, and F_z=T force.

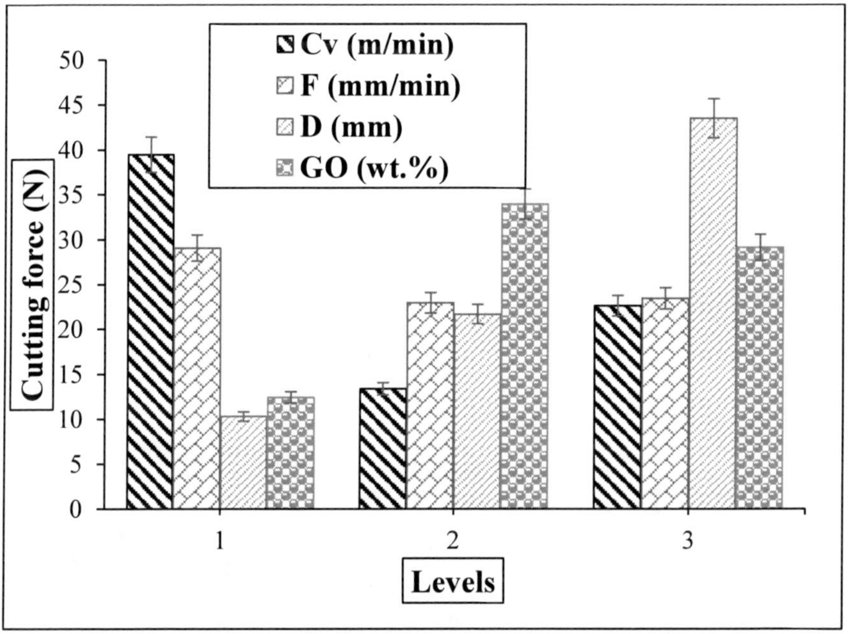

Figure 3.22. Cf Vs. Milling factors analysis (J. Kumar et al., 2022).

The fundamental determinant of the finish quality of a machined surface is the Cfs required. It is generally drastically diminished by increasing the cutting speed (Cv), the cutting depth, and F under control. Due to the complex structure of LPNC, Cf conditions for CFRP can vary significantly across a wide range of F. LPNC material showed different Cf characteristics at low, medium, and high cutting speeds. At low cutting speeds, hybrid materials have brittle fracture behaviour, which impacts machinability. Under high cutting speeds, the fibre encounters a high strain rate, leading to brittle failure at a lower strain (F. ji Wang et al., 2019). Since higher cutting speeds resulted in a greater material flow during machining when compared to medium cutting speeds, medium speeds are preferred for reduced Cfs (Palanikumar et al., 2006). Figure 3.22 plots the Cfs against the process variables. With a cutting speed of up to 25.12 m/min and an F of 240 mm/min, the Cf is determined to be at its lowest. At large F, the ploughing effect reduces the Cf (Palanikumar et al., 2006). By increasing the cutting depth, it is possible to enhance the Cf, as the cutting tool would penetrate deeper and remove more material at a faster rate (Mohammad Shahadath Hossain, 2019). The results show that the Cf drops significantly as the cutting speed increases to 25.12 m/min. At a lower depth of cut and wt.% of CNM, the Cf is significantly lowest.

3.3.3. Surface Roughness

Taylor Hobson Surtronic, Model No. S-128 measured Ra (Figure 3.23). An equipment that evaluates surface evenness has a stylus that can be traversed within the surface and convert perpendicular motion into an electrical signal. The filtered signal was used to calculate the roughness (Ra), which was then shown on a digital meter. The Ra quality attributes are analyzed by calculating the arithmetic means of roughness. The monitoring stylus has moved in several locations throughout the machine surface (Torres et al., 2017). The microscope image of the milling slot is displayed in Figure 3.24.

While milling LPNC material surfaces, cutting speeds and F are more critical than milling traditional materials. Numerous studies showed that raising the cutting temperature during milling causes the polymer to soften. Composite epoxy deterioration can occur when temperatures are too high, which can affect the quality of operations. The result of this phenomenon is that fibers have poor adhesive function (H. Wang et al., 2016). Additionally, a slightly higher Ra can be seen at a high feed level. The F and cut depth in the machine tool create vibrations that impact the milling operation's efficiency and surface quality. However, it would be assumed that the surface quality was lowered due to the increased depth of cut, which also caused an increase in Cf and higher TWR (Mohammad Shahadath Hossain, 2019). The higher F because of the heterogeneities of the composites and the material in polymeric materials cause higher vibrations during cutting. Figure 3.25 shows the plots for the variation of process parameters and their effect on Ra.

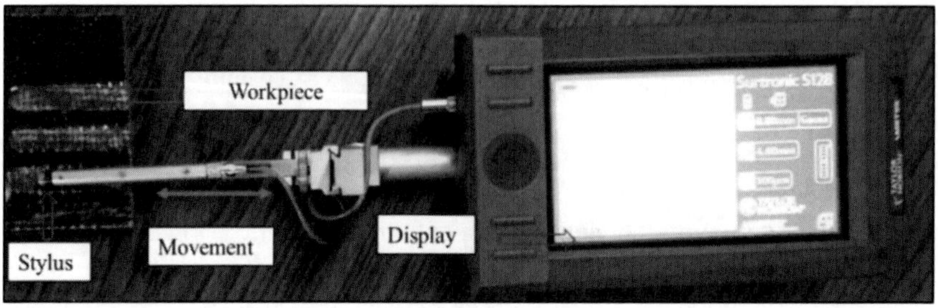

Figure 3.23. Ra setup (J. Kumar, Verma, Mondal, et al., 2021).

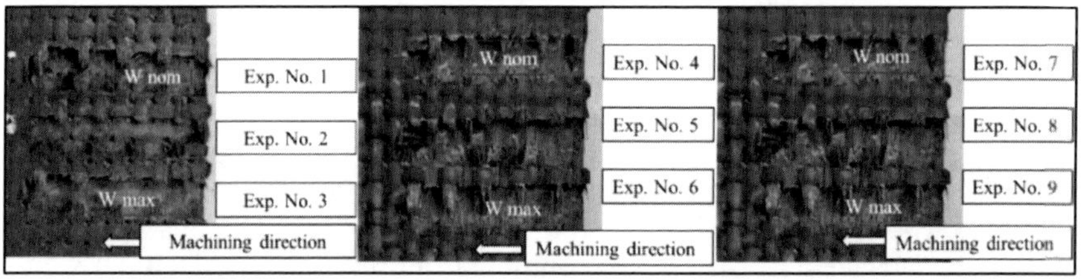

Figure 3.24. Milling-machined samples (J. Kumar & Verma, 2021b).

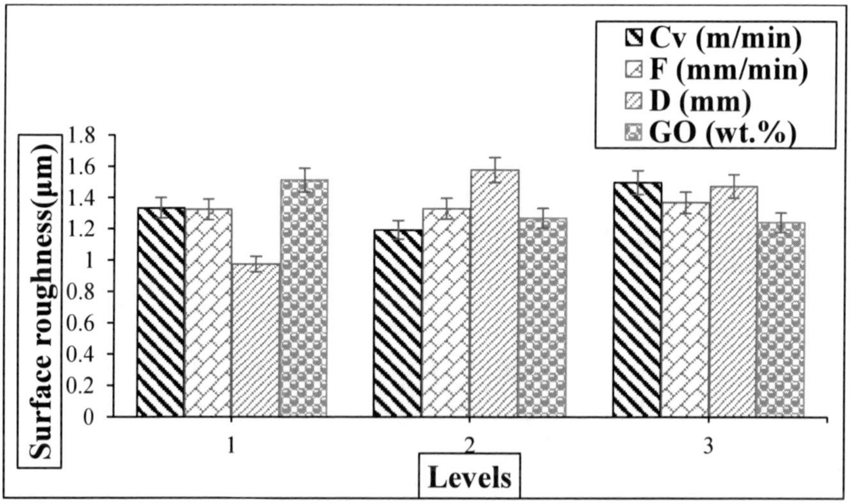

Figure 3.25. Ra Vs. Milling factors analysis (J. Kumar et al., 2022).

It has been noted that at a lower value depth of cut and a cutting speed of 25.12 m/min, an acceptable machined surface is obtained. It is recommended to use medium cutting and lower F while processing FRP composites for reduced Ra (Palanikumar, 2012). The research demonstrates that large aspect ratios affect all three wt.% of graphene nanomaterials. The presence of graphene nanomaterial causes the tool-workpiece interface to interact with lubrication. It shows that the polymer undergoes an increase in ductile transition for better machinability. It demonstrates that for lower cutting depths, higher graphene weight %, and F, the Ra lowers significantly with a 25.12 m/min cutting speed.

3.3.4. Delamination Factor

The analytical software ULTRA-CMOS5100 is utilized by the computerized tomography apparatus microscope (Model No. ZSM3780 T2) to evaluate Fd, as illustrated in Figure 3.26. Each specimen was examined for a single dataset thrice during the computerized microscopy process (Figure 3.27). Fd is calculated as the proportion of the nominal slot diameter to the maximum damaged slot length of the machined region. The milling-induced Fd is calculated using the following formula.

$$Delamination\ factor\ (F_d) = \frac{Maximum\ damage\ slot\ width\ (W_{max})}{Nominal\ damage\ slot\ width\ (W)} \quad (5.2)$$

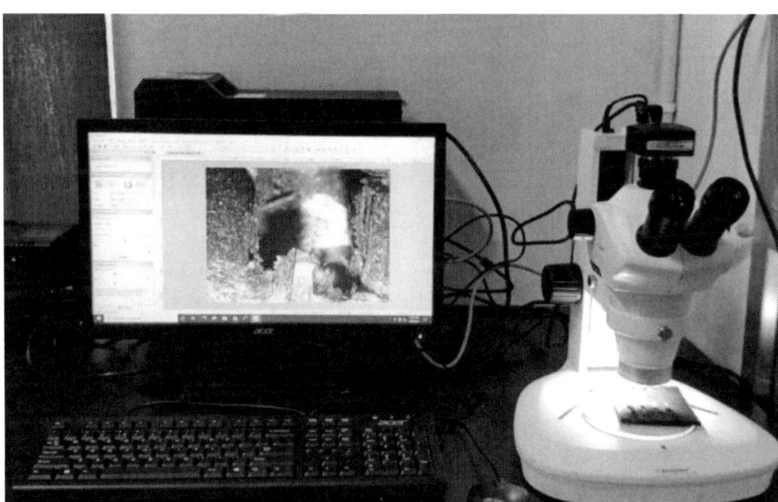

Figure 3.26. Fd measurement setup for milling process (J. Kumar & Verma, 2021b).

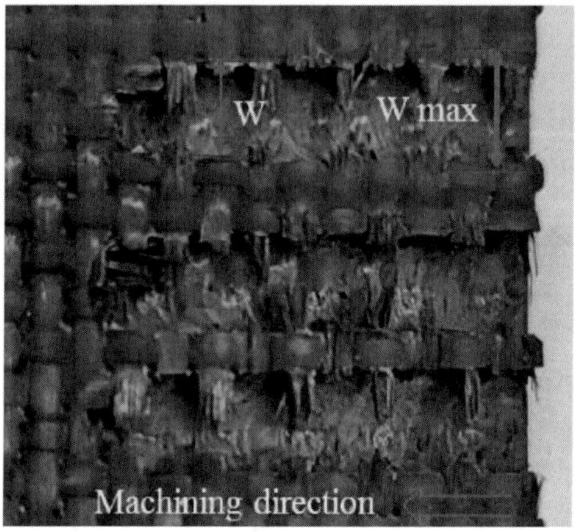

Figure 3.27. Milling sample with measurement of F_d (J. Kumar & Verma, 2021b).

Fd is a severe and challenging machining issue in PCs that occurs during the milling of fiber laminates. Nine experiments (Table 3.8) use different factor combinations and matching responses.

Table 3.8. Observation of milling-induced Fd (J. Kumar & Verma, 2021b)

Exp. No.	Process parameter				Delamination
	V_c (m/min)	F (mm/min)	D (mm)	GO (wt.%)	Fd
1	12.56	80	0.5	1	1.19
2	12.56	160	1	2	1.61
3	12.56	240	1.5	3	1.86
4	25.12	80	1	3	1.52
5	25.12	160	1.5	1	1.36
6	25.12	240	0.5	2	1.79
7	37.68	80	1.5	2	1.28
8	37.68	160	0.5	3	1.31
9	37.68	240	1	1	1.63

It harms manufacturing output, efficiency, and productivity when placing products. This type of damage caused by cutting tools can be avoided by carefully managing the cutting environment. The peak of the Cf assessed by the digital toolbox dynamometer occurs when the force is at its highest during the machining operation. Chip formation in metallic materials can reveal the deformation features of the milling process, but in LPNC, fibre and epoxy have different qualities, and damage development occurs in different laminates. Due to the characteristics and weaves of plain-woven fibre, the Fd zone in the suggested nanocomposite milling test becomes more complicated. The level of Cf varies because of these characteristics. Chip fragmentation occurs at high strain rates during milling of modified LNPC, and early resin fracture is indicated by high F and cutting speeds. However, with a more remarkable combination of feed and cutting speed, the heat produced might raise the temperature of the cutting zone. As a result, the material's ductility is increased by lengthening the molecular chain with long-range mobility. As a result, there is a tendency for fibres to fracture more frequently, leading to more severe damage (F. ji Wang et al., 2019). The longer continuous carbon fibre strands instantly fracture at the higher cutting speed and F, highlighting the chip microstructure and revealing significant fibre crack growth on the chip at irregular extents. It was observed (Figure 3.28) that the binding strength between the fibre and matrix reduced due to increased rubbing at the work tool boundary area at high cutting speed and feed, causing the fibres to either peel (Fd) or be separated from the polymer. It is noticed that the Fd factor significantly reduces with cutting speed up to 37.68 m/min. A lower depth of cut and wt.% of CNMs significantly lowers the Fd.

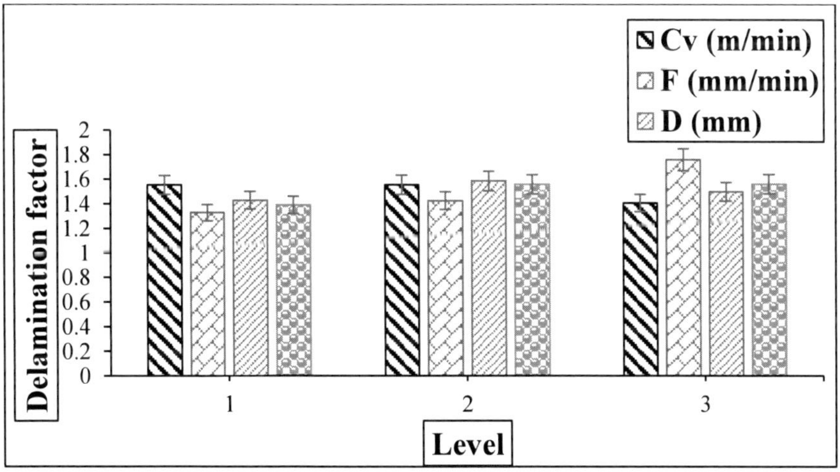

Figure 3.28. Fd Vs. Milling factors analysis (J. Kumar et al., 2022).

Conclusion

The machinability aspect and influence of process constraints on the drilling and milling performances have been investigated in this chapter. When subjected to a real machining test, the prime objective is to identify the nanocomposites' machining capacity. This chapter offers the experimental investigation under varying cutting conditions and three different ranges of modified epoxy by CNMs effect onto laminated PCs. The impact of varying cutting settings on the Th, T, MRR, Cf, and Ra have been investigated to calculate the machinability characteristics of the polymers. Based on the findings, the conclusion can be outlined as follows:

- This study illustrates the analysis of force signals produced under various machining parameters. Five steps, representing the various drilling tool-workpiece interactions, are indicated by the whole drilling force signal and cutting parameters examined.
- For drilling LPNC material, the combined impact of a lower F and greater S could achieve better surface quality. The lower plastic transition temperature causes the increased cutting speed to provide less Th and T.
- The proposed laminate nanocomposites Cf and MRR trends are comparable to those of conventional materials. Since a medium cutting speed is combined with a high F, the MRR value increases.
- By altering the cutting speed, F, depth of cut, and the amount of added nanomaterial, it is possible to regulate the machined surface quality.

The varying parameters effect shows the application potential of the composite for multifunction performance components subjected to continuous stress and strain. Additionally, the machined surface quality at a medium cutting speed demonstrates the better application possibilities for an efficient Milling test. During the drilling operation, the control of process constraints plays a significant role in creating a defect-free machined hole. The next chapter involves the experimental investigation of unconventional machining during laminate manufacturing. Hence, an effort is planned to maintain an efficient machining environment to tackle complex real-world issues during LPNC machining.

References

Abrão, A. M., Rubio, J. C. C., Faria, P. E., & Davim, J. P. (2008). The effect of cutting tool geometry on thrust force and delamination when drilling glass fibre reinforced plastic composite. *Materials and Design*, *29*(2), 508–513.

Adak, N. C., Chhetri, S., Murmu, N. C., Samanta, P., Kuila, T., & Lee, J. H. (2018). Experimental and numerical investigation on the mechanical characteristics of polyethylenimine functionalized graphene oxide incorporated woven carbon fibre/epoxy composites. *Composites Part B: Engineering*, *156*, 240–251. https://doi.org/10.1016/j.compositesb.2018.08.086.

Alagumurthi, N., Palanikumar, K., Elansezhian, R., & Anand, G. (2017). Delamination study on newly formulated Ni-P coated glass fibre/nanowire - reinforced polymer composite using Grey relational analysis. *International Journal of Additive and Subtractive Materials Manufacturing*, *1*(3/4), 372.

Azmi, A. I., Lin, R. J. T., & Bhattacharyya, D. (2013). Machinability study of glass fibre-reinforced polymer composites during end milling. *International Journal of Advanced Manufacturing Technology*, *64*(1–4), 247–261.

Bhat, R., Mohan, N., Sharma, S., Agarwal, R. A., Rathi, A., & Subudhi, K. A. (2019). Multi-response optimization of the thrust force, torque and surface roughness in drilling of glass fiber reinforced polyester composite using GRA-RSM. *Materials Today: Proceedings*, *19*(2), 333–338.

Boughdiri, I., Giasin, K., Mabrouki, T., & Zitoune, R. (2021). Effect of cutting parameters on thrust force, torque, hole quality and dust generation during drilling of GLARE 2B laminates. *Composite Structures*, *261*(January), 113562. https://doi.org/10.1016/j.compstruct.2021.113562.

Brinksmeier, E. (1990). Prediction of Tool Fracture in Drilling. *CIRP Annals - Manufacturing Technology*, *39*(1), 97–100.

Caggiano, A. (2018). Machining of fibre reinforced plastic composite materials. *Materials*, *11*(3), 442 (1-20). https://doi.org/10.3390/ma11030442.

Caggiano, A., Rimpault, X., Teti, R., Balazinski, M., Chatelain, J. F., & Nele, L. (2018). Machine learning approach based on fractal analysis for optimal tool life exploitation in CFRP composite drilling for aeronautical assembly. *CIRP Annals*, *67*(1), 483–486.

Dong Wang, G., Li, N., Kirwa Melly, S., Peng, T., chi Li, Y., Di Zhao, Q., & De Ji, S. (2019a). Monitoring the drilling process of GFRP laminates with carbon nanotube buckypaper sensor. *Composite Structures*, *208*, 114–126.

Dong Wang, G., Li, N., Kirwa Melly, S., Peng, T., chi Li, Y., Di Zhao, Q., & De Ji, S. (2019b). Monitoring the drilling process of GFRP laminates with carbon nanotube buckypaper sensor. *Composite Structures*, *208*, 114–126.

El-Sonbaty, I., Khashaba, U. A., & Machaly, T. (2004). Factors affecting the machinability of GFR/epoxy composites. *Composite Structures*, *63*(3–4), 329–338.

Fu, J., Zhang, M., Jin, L., Liu, L., Li, N., Shang, L., Li, M., Xiao, L., & Ao, Y. (2019). Enhancing interfacial properties of carbon fibers reinforced epoxy composites via Layer-by-Layer self assembly GO/SiO 2 multilayers films on carbon fibers surface. *Applied Surface Science*, *470*, 543–554.

Gaitonde, V. N., Karnik, S. R., Rubio, J. C., Correia, A. E., Abrão, A. M., & Davim, J. P. (2008). Analysis of parametric influence on delamination in high-speed drilling of carbon fiber reinforced plastic composites. *Journal of Materials Processing Technology*, *203*(1–3), 431–438.

Gupta, M., & Kumar, S. (2015). Investigation of surface roughness and MRR for turning of UD-GFRP using PCA and Taguchi method. *Engineering Science and Technology, an International Journal*, *18*(1), 70–81.

Ha, S. J., Kim, K. B., Yang, J. K., & Cho, M. W. (2017). Influence of cutting temperature on carbon fiber-reinforced plastic composites in high-speed machining. *Journal of Mechanical Science and Technology*, *31*(4), 1861–1867.

He, R., Chang, Q., Huang, X., & Bo, J. (2018). Improved mechanical properties of carbon fiber reinforced PTFE composites by growing graphene oxide on carbon fiber surface. *Composite Interfaces*, *25*(11), 995–1004. https://doi.org/10.1080/09276440.2018.1451677.

Hocheng, H., & Tsao, C. C. (2003). Comprehensive analysis of delamination in drilling of composite materials with various drill bits. *Journal of Materials Processing Technology*, *140*, 335–339.

Isbilir, O., & Ghassemieh, E. (2012). Delamination and wear in drilling of carbon-fiber reinforced plastic composites using multilayer TiAlN/TiN PVD-coated tungsten carbide tools. *Journal of Reinforced Plastics and Composites*, *31*(10), 717–727.

Janakiraman, A., Pemmasani, S., Sheth, S., Kannan, C., & Balan, A. S. S. (2020). Experimental Investigation and Parametric Optimization on Hole Quality Assessment During Drilling of CFRP/GFRP/Al Stacks. *Journal of The Institution of Engineers (India): Series C*, *101*(2), 291–302.

Jenarthanan, M. P., & Jeyapaul, R. (2018). Optimisation of machining parameters on milling of GFRP composites by desirability function analysis using Taguchi method. *International Journal of Engineering, Science and Technology*, *5*(4), 22.

Kaybal, H. B., Ünüvar, A., Koyunbakan, M., & Avcı, A. (2019). A novelty optimization approach for drilling of CFRP nanocomposite laminates. *International Journal of Advanced Manufacturing Technology*, *100*(9–12), 2995–3012.

Komanduri, R. (1997). Machining of fiber-reinforced composites. *Machining Science and Technology*, *1*(1), 113–152.

Krishnamoorthy, A., Rajendra Boopathy, S., Palanikumar, K., & Paulo Davim, J. (2012). Application of grey fuzzy logic for the optimization of drilling parameters for CFRP composites with multiple performance characteristics. *Measurement: Journal of the International Measurement Confederation*, *45*(5), 1286–1296.

Kumar, D., & Gururaja, S. (2020). Machining damage and surface integrity evaluation during milling of UD-CFRP laminates: Dry vs. cryogenic. *Composite Structures*, *247*, 112504.

Kumar, D., & Singh, K. K. (2015). An approach towards damage free machining of CFRP and GFRP composite material: A review. *Advanced Composite Materials*, *24*, 49–63.

Kumar, D., & Singh, K. K. (2019). Investigation of delamination and surface quality of machined holes in drilling of multiwalled carbon nanotube doped epoxy/carbon fiber reinforced polymer nanocomposite. *Proceedings of the Institution of Mechanical Engineers, Part L: Journal of Materials: Design and Applications*, *233*(4), 647–663.

Kumar, J., Abhishek, K., Xu, J., & Verma, R. K. (2022). Experimental Investigation on Machine-Induced Damages during the Milling Test of Graphene/Carbon Incorporated Thermoset Polymer Nanocomposites. *Journal of Composites Science*, *6*(3), 1–12. https://doi.org/10.3390/jcs6030077.

Kumar, J., & Kumar Verma, R. (2021). A novel methodology of Combined Compromise Solution and Principal Component Analysis (CoCoSo-PCA) for machinability investigation of graphene nanocomposites. *CIRP Journal of Manufacturing Science and Technology*, *33*, 143–157. https://doi.org/https://doi.org/10.1016/j.cirpj.2021.03.007.

Kumar, J., Kumar Verma, R., & Debnath, K. (2020). A new approach to control the delamination and thrust force during drilling of polymer nanocomposites reinforced by graphene oxide/carbon fiber. *Composite Structures*, *253*(August), 112786. https://doi.org/10.1016/j.compstruct.2020.112786.

Kumar, J., & Verma, R. K. (2021a). A new criterion for drilling machinability evaluation of nanocomposites modified by graphene/carbon fiber epoxy matrix and optimization using combined compromise solution. *Surface Review and Letters*, *28*(9). https://doi.org/10.1142/S0218625X21500827.

Kumar, J., & Verma, R. K. (2021b). Delamination assessment during machining of laminated polymer nanocomposite. *International Journal of Modern Manufacturing Technologies*, *13*(2), 109–115. https://doi.org/10.54684/ijmmt.2021.13.2.109.

Kumar, J., & Verma, R. K. (2021c). Experimental investigation for machinability aspects of graphene oxide/carbon fiber reinforced polymer nanocomposites and predictive modeling using hybrid approach. *Defence Technology*, *17*(5), 1671–1686. https://doi.org/10.1016/j.dt.2020.09.009.

Kumar, J., & Verma, R. K. (2023). An integrated module for predictive modelling and machinability appraisal during milling of modified graphene/epoxy nanocomposites. *Australian Journal of Mechanical Engineering*, *21*(2), 682–694. https://doi.org/10.1080/14484846.2021.1913854.

Kumar, J., & Verma, R. K. (2024). A Bionic Inspired Dragonfly Algorithm for Parametric Optimization and Damage Investigation During Machining of Modified Polymer Nanocomposite. *Surface Review and Letters*, *April*, 25500532. https://doi.org/10.1142/s0218625x25500532.

Kumar, J., Verma, R. K., & Mondal, A. K. (2021). Taguchi- Grey Theory Based Harmony Search Algorithm (GR-HSA) for Predictive Modeling and Multi-Objective Optimization in Drilling of Polymer Composites. *Experimental Techniques*, *45*(4), 531–548. https://doi.org/10.1007/s40799-020-00428-y.

Kumar, J., Verma, R. K., Mondal, A. K., & Singh, V. K. (2021). A hybrid optimization technique to control the machining performance of graphene/carbon/polymer (epoxy) nanocomposites. *Polymers and Polymer Composites*, *29*(9), S1168–S1180. https://doi.org/10.1177/09673911211046789.

Kurniawan, R., Thirumalai Kumaran, S., Arumuga Prabu, V., Zhen, Y., Park, K. M., Kwak, Y. I., Mofizul Islam, M., & Ko, T. J. (2017). Measurement of burr removal rate and analysis of machining parameters in ultrasonic assisted dry EDM (US-EDM) for deburring drilled holes in CFRP composite. *Measurement: Journal of the International Measurement Confederation*, *110*, 98–115.

Li, L., Yan, C., Xu, H., Liu, D., Shi, P., Zhu, Y., Chen, G., Wu, X., & Liu, W. (2019). Improving the interfacial properties of carbon fiber–epoxy resin composites with a graphene-modified sizing agent. *Journal of Applied Polymer Science*, *136*(9), 1–10.

Li, P., Qiu, X., Li, C., Niu, Q., Chen, A., & Ko, T. J. (2019). Hole exit damage and tool wear during the drilling of CFRP with a double-point angle drill. *Journal of Mechanical Science and Technology*, *33*(5), 2363–2370.

Liu, D. F., Tang, Y. J., & Cong, W. L. (2012). A review of mechanical drilling for composite laminates. *Composite Structures*, *94*(4), 1265–1279. https://doi.org/10.1016/j.compstruct.2011.11.024.

Meinhard, D., Haeger, A., & Knoblauch, V. (2021). Drilling induced defects on carbon fiber-reinforced thermoplastic polyamide and their effect on mechanical properties. *Composite Structures*, *256*(October 2020), 113138. https://doi.org/10.1016/j.compstruct.2020.113138.

Miyake, T., Mukae, K., & Futamura, M. (2016). Evaluation of machining damage around drilled holes in a CFRP by fiber residual stresses measured using micro-Raman spectroscopy. *Mechanical Engineering Journal*, *3*(6), 16–00301.

Mohammad Shahadath Hossain, N. R. D. (2019). Machinability Study of a Kevlar and Glass Reinforced Polyester. *Journal of Mechanical Engineering*, *49*(1), 18–26.

Mohan, N. S., Ramachandra, A., & Kulkarni, S. M. (2005). Influence of process parameters on cutting force and torque during drilling of glass-fiber polyester reinforced composites. *Composite Structures*, *71*(3–4), 407–413. https://doi.org/10.1016/j.compstruct.2005.09.039.

Ogawa, K., Aoyama, E., Inoue, H., Hirogaki, T., Nobe, H., Kitahara, Y., Katayama, T., & Gunjima, M. (1997). Investigation on cutting mechanism in small diameter drilling for GFRP (thrust force and surface roughness at drilled hole wall). *Composite Structures*, *38*(1–4), 343–350.

Palanikumar, K. (2012). Analyzing surface quality in machined composites. In *Machining Technology for Composite Materials*. Woodhead Publishing Limited.

Palanikumar, K., Karunamoorthy, L., & Karthikeyan, R. (2006). Multiple performance optimization of machining parameters on the machining of GFRP composites using carbide (K10) tool. *Materials and Manufacturing Processes*, *21*(8), 846–852.

Pang, J. S., Ansari, M. N. M., Zaroog, O. S., Ali, M. H., & Sapuan, S. M. (2014). Taguchi design optimization of machining parameters on the CNC end milling process of halloysite nanotube with aluminium reinforced epoxy matrix (HNT/Al/Ep) hybrid composite. *HBRC Journal*, *10*(2), 138–144.

Papageorgiou, D. G., Li, Z., Liu, M., Kinloch, I. A., & Young, R. J. (2020). Mechanisms of mechanical reinforcement by graphene and carbon nanotubes in polymer nanocomposites. *Nanoscale*, *12*(4), 2228–2267. https://doi.org/10.1039/c9nr06952f.

Park, S. Y., Choi, W. J., Choi, C. H., & Choi, H. S. (2018). Effect of drilling parameters on hole quality and delamination of hybrid GLARE laminate. *Composite Structures*, *185*, 684–698.

Ponappa, K., Sasikumar, K. S. K., Sambathkumar, M., & Udhayakumar, M. (2019). Multi-objective optimization of edm process parameters for machining of hybrid aluminum metal matrix composites (Al7075/TiC/B4C) using genetic algorithm. *Surface Review and Letters*, *26*(10), 1950071.

Qin, W., Vautard, F., Drzal, L. T., & Yu, J. (2015). Mechanical and electrical properties of carbon fiber composites with incorporation of graphene nanoplatelets at the fiber-matrix interphase. *Composites Part B: Engineering*, *69*, 335–341.

Rajagopal, K., & Anand, S. (1999). Assessment of circularity error using a selective data partition approach. *International Journal of Production Research*, *37*(17), 3959–3979.

Rao, G. V. G., Mahajan, P., & Bhatnagar, N. (2007). Micro-mechanical modeling of machining of FRP composites - Cutting force analysis. *Composites Science and Technology*, *67*(3–4), 579–593.

Ray, D., Puri, A. B., & Nagahanumaiah. (2019). Investigation on Cutting Forces and Surface Finish in Mechanical Micro Milling of Zr-Based Bulk Metallic Glass. *Journal of Advanced Manufacturing Systems*, *18*(1), 113–132.

Samuel, G. L., & Shunmugam, M. S. (2003). Evaluation of circularity and sphericity from coordinate measurement data. *Journal of Materials Processing Technology*, *139*(1–3), 90–95.

Sanjay, M. R., & Yogesha, B. (2018). Studies on hybridization effect of jute/kenaf/E-glass woven fabric epoxy composites for potential applications: Effect of laminate stacking sequences. *Journal of Industrial Textiles*, *47*(7), 1830–1848.

Saoudi, J., Zitoune, R., Mezlini, S., Gururaja, S., & Seitier, P. (2016). Critical thrust force predictions during drilling: Analytical modeling and X-ray tomography quantification. *Composite Structures*, *153*, 886–894.

Shunmugesh, K., Akhil, K. T., Aravind, S., & Pramodkumar, M. (2017). Optimization of Drilling Characteristics using Grey-Fuzzy logic in Glass Fiber Reinforced Polymer (GFRP). *Materials Today: Proceedings*, *4*(8), 8938–8947.

Sorrentino, L., Turchetta, S., & Bellini, C. (2017). In process monitoring of cutting temperature during the drilling of FRP laminate. *Composite Structures*, *168*, 549–561.

Sridharan, V., Raja, T., & Muthukrishnan, N. (2016). Study of the Effect of Matrix, Fibre Treatment and Graphene on Delamination by Drilling Jute/Epoxy Nanohybrid Composite. *Arabian Journal for Science and Engineering*, *41*(5), 1883–1894. https://doi.org/10.1007/s13369-015-2005-2.

Srivastava, V. K., Jain, P. K., Kumar, P., Pegoretti, A., & Bowen, C. R. (2020). Smart Manufacturing Process of Carbon-Based Low-Dimensional Structures and Fiber-Reinforced Polymer Composites for Engineering Applications. *Journal of Materials Engineering and Performance*, *29*(7), 4162–4186.

Torres, A., Luis, C. J., & Puertas, I. (2017). EDM machinability and surface roughness analysis of TiB2using copper electrodes. *Journal of Alloys and Compounds*, *690*, 337–347.

Tsao, C. C. (2008). Thrust force and delamination of core-saw drill during drilling of carbon fiber reinforced plastics (CFRP). *International Journal of Advanced Manufacturing Technology*, *37*(1–2), 23–28.

Vijayan, D., & Rajmohan, T. (2019). Modeling and evolutionary computation on drilling of carbon fiber-reinforced polymer nanocomposite: an integrated approach using RSM based PSO. *Journal of the Brazilian Society of Mechanical Sciences and Engineering*, *41*(10), 1–17.

Wang, F. ji, Zhang, B. yu, Zhao, M., Cheng, D., & Wang, Z. gang. (2019). Evolution laws of fiber-matrix interface cracks in machining of carbon fiber reinforced polymer. *International Journal of Advanced Manufacturing Technology*, *101*(1–4), 963–977.

Wang, H., Sun, J., Zhang, D., Guo, K., & Li, J. (2016). The effect of cutting temperature in milling of carbon fiber reinforced polymer composites. *Composites Part A: Applied Science and Manufacturing*, *91*, 380–387.

Wang, M., Cheraghi, S. H., & Masud, A. S. M. (1999). Circularity error evaluation theory and algorithm. *Precision Engineering*, *23*(3), 164–176.

Wei, Y., An, Q., Ming, W., & Chen, M. (2016). Effect of drilling parameters and tool geometry on drilling performance in drilling carbon fiber-reinforced plastic/titanium alloy stacks. *Advances in Mechanical Engineering*, *8*(9), 1–16.

Wu, C., Chen, Z., Wang, F., Hu, Y., Wang, E., Rao, Z., & Zhang, X. (2019). Mechanical and drilling properties of graphene oxide modified urea-melamine-phenol formaldehyde composites reinforced by glass fiber. *Composites Part B: Engineering*, *162*, 378–387.

Xu, J., Li, C., Chen, M., Mansori, M. El, & Davim, J. P. (2019). On the analysis of temperatures , surface morphologies and tool wear in drilling CFRP / Ti6Al4V stacks under different cutting sequence strategies. *Composite Structures*, *234*, 111708.

Zarif Karimi, N., Heidary, H., Fotouhi, M., & Minak, G. (2017). Experimental analysis of GFRP laminates subjected to compression after drilling. *Composite Structures*, *169*, 144–152.

Zitoune, R., Krishnaraj, V., Collombet, F., & Le Roux, S. (2016). Experimental and numerical analysis on drilling of carbon fibre reinforced plastic and aluminium stacks. *Composite Structures*, *146*, 148–158.

Chapter 4

Defects and Damage Generation Mechanisms During Machining

4.1. Challenges During the Machining of Laminated Polymer Nanocomposites

Aeronautical, naval, space and automotive components have used laminate polymer composites (LPC) for their improved mechanical qualities and fatigue life for two decades. Apart from metal alloys and non-metallic materials, it can be used instead of common engineering materials. The physical and structural properties of the chosen manufacturing materials must be distinctive, characterized by a high resistance to degradation and a low specific weight. It can ensure economic performance and safety factors in these multifunctional conditions. Cross-linked connections, in addition to thermoset polymers, enhance the thermal, mechanical, and tensile properties. LPCs have unique machining challenges because of their cross-linking, anisotropy, and non-homogeneity (Altin Karataş & Gökkaya, 2018; Geng et al., 2019). Recent research initiatives have focused on studies about machining laminated polymer nanocomposites (LPNC). Manufacturing organizations, including those involved in the assembly of automobiles, batteries, spacecraft, turbines, etc., face a formidable challenge when machining nanocomposite materials. Manufacturers encountered numerous complicated issues during the machining of LPNC, such as circularity error (Ce), delamination factor (Fd), burr, fibre pull, matrix debonding, and so on (Gara et al., 2018; He et al., 2017; Stone & Krishnamurthy, 1996).

A combination of factors, including complicated materials, tool geometry, and environmental conditions beyond the stipulated criteria and uncontrolled process parameters during the machining test, can cause significant damage and eventual material failure. To prevent the engineering component from failing, research is being done on the drilling/milling method for nanocomposite, which requires highly productive, economical, and lasting materials. Several research studies are conducted for a delamination-induced drilling/milling operation, considering the parameters of spindle speed (S), feed (F), depth of cut (D), tool shape, tool materials, and tool coating.

However, these techniques have several drawbacks, including layered nanomaterial machining. The percentage of nanomaterials is thought to be a process limitation for the milling process's control Fd (Heidary et al., 2018; X. Wang et al., 2020). Many eminent researchers have been looking into strategies to control delamination damage in laminates reinforced with macro fibres (glass, carbon, aramid, etc.). Machining indices are among the process parameters controlled by a practical statistical module and optimization tools (Ray et al., 2019).

The quality of the produced components and machining efficiency are greatly influenced by controllable (S, F, D, tool diameter, etc.) and noncontrollable (operating environment) elements during laminate nano/composites machining. Variations in process restrictions, material quality, and irregular cutting force all contribute to the damage and faults. The term "composite damage" refers to the interplay/layers' failure tendency that results from the exceptionally severe perforation of composite laminates, which can irregularly create unforeseen damages and substantial assembly failure (Eneyew & Ramulu, 2014; Gemi et al., 2019). According to experimental research, the machining of fiber-reinforced polymer (FRP) composites and the occurrence of induced defects are directly affected by the input variables. These variables include S, F, cutting angle, tool sizes, tool materials, and operation mode (Sorrentino et al., 2018). Composites often go through several machining operations before they are found to be used in fundamental engineering applications. The drilling and milling processes are essential for hole and slot development to install different structures or components. Various industries, such as the automotive, aerospace, and marine sectors, have already begun utilizing nanocomposites in their structures (Sawangphruk et al., 2013). One of the most common issues with milling laminate polymers is damage to the composite, which is commonly referred to as delamination. This damage is an example of drilling-induced damage in image processing. Delamination failures are a significant and unforeseen consequence of drilling and milling polymer laminates (John et al., 2019). The anisotropic nature and multiple-phase structure of composites cause them to be irregular, interrupting the milling and drilling processes. Damage to the inlet/outlet section from uncut composite laminates during drilling could lead to delamination. This phenomenon is observed at the perimeter of the cut that is produced in the laminates during milling. Drilling and milling during the machining of polymers can also seriously harm subsurface deformation and delamination (Gara et al., 2018; Xu & El Mansori, 2016). Substructure defects include fiber pullouts, spalling, interfacial debonding, and matrix deformation and splitting (Zhang et al., 2019). Therefore, to increase product durability and efficiency, the material defects caused by delamination should be fixed using a suitable range of cutting constraints, tool geometries, tool sizes, and cutting conditions (Bhatnagar et al., 2004; Hocheng & Tsao, 2003). However, fiber patterns and interfacial interactions with the fiber matrix are the primary factors determining structural strength (D. Kumar & Singh, 2019). This section examines the effects of process parameters on machining damage control and the defects and damage analysis of LPNC during the drilling and milling process.

Because polymer fibers behave in an anisotropic and abrasive manner, their machining mechanism differs entirely from that of metallic materials. Controlling burr formation, fiber pullout, splintering, delamination, or cracking is crucial. The homogenous removal of metals is not the same as the machining trend of polymeric material (Liu et al., 2012; Voß et al., 2014). The various characteristics of the fiber and matrix significantly impact processing according to fiber orientation. It is mainly caused by both soft matrix and stiff abrasive fibers; it is challenging to select the correct cutting conditions (Anarghya et al., 2018; Gaugel et al., 2016). Observational results show that composite materials have a 10–100 times poorer crack resistance and are less susceptible to plastic distortion than regular steels (Adak et al., 2019). Customer requests have focused on producing modified machining technology, which has had strong production handling challenges in recent years. One significant benefit of composites is their ability to make structures near to design specifications. Therefore, to attain high productivity and reduced costs, a novel approach to polymer machining is required (Liu et al., 2012). Parts production and assembly are two manufacturing processes that rely on machining. Rapid deterioration of structural efficiency and shoddy assembly result from subpar privilege machining (Ho-Cheng & Dharan, 1990; Khashaba, 2004). The necessity for mechanical parts and assemblies to be riveted or bolted makes hole processing a common approach for secondary composite machining (Bielawski et al., 2017). Multiple procedures are examined, including computer numerical control (CNC) drilling, ultrasonic drilling, laser beam drilling, etc (Dahnel et al., 2016; Santhanakrishnan Balakrishnan et al., 2019). The conventional drilling method with a twist drill is often the most cost-effective and efficient when working with composites. A medium-sized industrial report found that up to 40% of the process is devoted to hole creation, while the drilling technique receives insufficient attention. Poorly machined quality accounts for almost 60% of all part rejections (Figures 4.1 and 4.2), leading to high costs and subpar products.

Nevertheless, the twist drill's geometry is relatively complex when contrasted with that of a straight-edge drill. The efficacy of the drill varies; the outer diameter is the most effective, while the central area is the least efficient (Hocheng & Tsao, 2003; Liu et al., 2012). The chisel edge has a negative rake angle, and the lips are close to the twist drill center. A significant negative rake angle will enhance this process and increase the difficulty of chip production. As it gets closer to the drill center, the relative speed decreases linearly and eventually approach zero. Consequently, this limits its ability to create holes. Therefore, materials are anticipated to extrude rather than cut when they pass through the drilling point's chisel edge and into the hole. The heat caused the chisel edge of the drill tool to become damaged, requiring a strong thrust to push the twist drill through the material (El-Sonbaty et al., 2004; Saoudi et al., 2016).

Composites often go through several machining operations before they are found to be used in fundamental engineering applications. Drilling is one of the most critical steps in developing holes to install frameworks or equipment. Various industries, such as the automotive, aerospace, and marine sectors, have already begun utilizing nanocomposites in their structures (Sawangphruk et al., 2013). (John et al., 2019). (Gara et al., 2018; Haeger, Schoen, et al., 2016). (Haeger, Meinhard, et al., 2016). (Bhatnagar et al., 2004; Hocheng & Tsao, 2003) (D. Kumar & Singh, 2019).

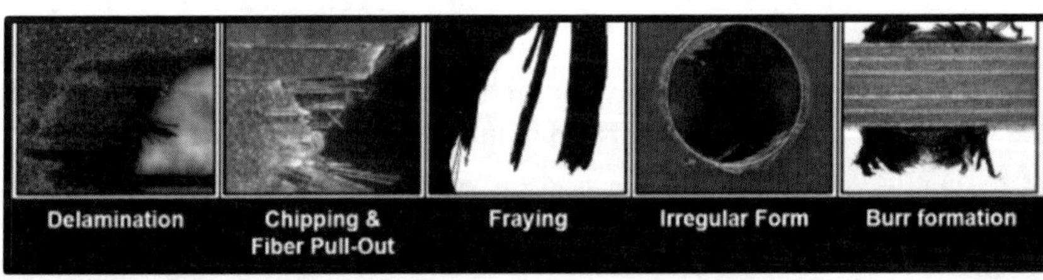

Figure 4.1. Drilling induced defects (Meinhard et al., 2021).

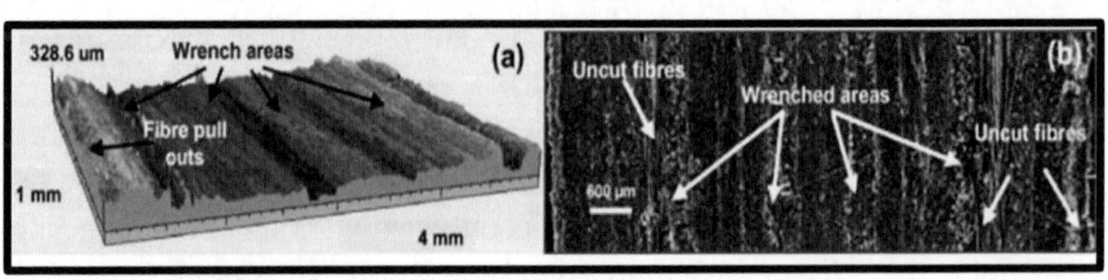

Figure 4.2. Milling-induced defects (Haddad et al., 2014).

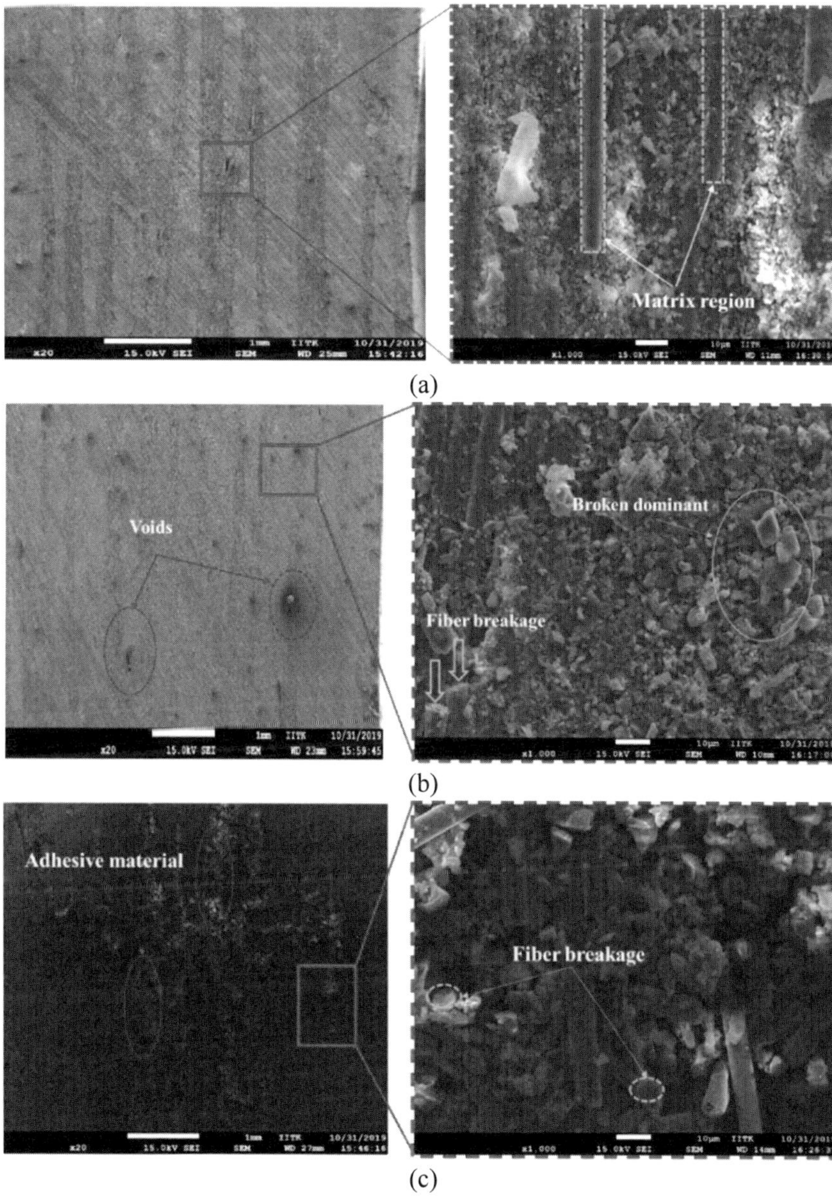

Figure 4.3. Post-drilling process at 2400 rpm, 80 mm/min (a) GO-1 wt.%; (b) GO-2 wt. %; (c) GO-3 wt. % (J. Kumar & Verma, 2021b).

4.2. Defects and Damage Generation Mechanisms

4.2.1. Investigation of Drilling-Induced Damages

Microstructure analysis confirmed surface characteristics and flaws on drilled samples at 1, 2, and 3% graphene oxide (GO) modified polymer. Figure 4.3 (a-c) displays FESEM images of graphene nanocomposite drilling flaws like fiber breaking, fiber pullout, deposition, and matrix cracking. There is negligible damage at this 1 weight percent GO concentration, as shown in Figure 4.3 (a). In addition, the matrix region has adequate bonding. The balanced addition of weight percent of GO to the epoxy matrix is primarily responsible for this (Dixit et al., 2018; Kostagiannakopoulou et al., 2017). Compared to micro or macro reinforcing agents, nanofillers can enhance the necessary mechanical qualities with a small quantity of content (Krishnamoorthy et al., 2012). The presence of GO enhances the mechanical characteristics of the polymer and reduces surface damage during machining by providing the necessary lubrication to the tool-chip interface. In Figure 4.3 (b), the indications of damage, including disintegration, are apparent. An additional consideration is attained to the machining-induced damages in Figure 4.3 (c), where the addition of GO 3 weight percent to the epoxy matrix restores the laminates' flexural strength. With a higher F, it takes more cutting force to pierce the laminates. An increase in thrust force causes a more significant damage zone from delamination because of the combined effects of increased F and cutting force. Delamination is widened by the increased F, mainly due to the drill bit's chisel edge. As with the cutting mechanism, the composite layers that come into direct contact with the chisel edge extrude the fastest because of the high F (Isbilir & Ghassemieh, 2012; Ragunath et al., 2017). Increased bending caused by the drilling force pushing the work material can result in matrix debonding, fiber pull/fracture, and crack initiation (Heidary et al., 2018; Palanikumar, 2006, 2011). Consequently, the optimal parametric configuration and balanced inclusion of nanomaterials are essential for a cost-effective machining environment.

4.2.2. Investigation of Milling-Induced Damages

Increased surface damage, segmentation, and wrinkles are observed in the microstructure images of the machined sample when the F is increased to 240 mm/min. More significant vibration during the cutting process caused by a high F increases the risk of fracture failure (Davim & Reis, 2005; Puw & Hocheng, 1993; Rajesh Mathivanan et al., 2016; Thakur et al., 2020). The increased F caused damage because the fiber/matrix breakdown reduced interfacial strength (Davim & Reis, 2005). Reduced failure, improved machinability, and flawless finishing are all signs that the segmentation pattern is adequate at the lower F and higher S. Even with a little reduction in F and an increase in S, a minor number of fracture propagations involving fiber pullout are noted in polymeric resin (Davim & Reis, 2005; Erkan et al., 2013; He et al., 2017; Sheikh-Ahmad et al., 2017). It may be achievable by reducing the cutting force to bend the cutting tool-less. The chip releases smoothly when the cutting chip rate exceeds the S and reduces uncut fibre yarn burrs (Azmi et al., 2013; Ha et al., 2017). Epoxy/CFRP composites have higher fibre pull with deeper cuts and more machining surface damage (F. ji Wang et al., 2019). Loaded nanofiller samples reduce damage from improved mechanical properties and microstructure (Saravanakumar et al., 2019). Consequently, the cutting force and graphene nanomaterials affected improved CFRPs with less fiber breakage and more uniform surface topography (Papadopoulos et al., 2016). Increased strain distribution and damage resistance result from adding carbon nanomaterials to the matrix and reinforcement's contact surface, which improves the high surface-to-volume ratio. (X. Wang et al., 2020). Previous studies have shown that composites supplemented with graphene oxide and carbon fibre can reduce delamination damage while maintaining a lower F (Razfar & Zadeh, 2009; X. Wang et al., 2020). The impact of machining on a carbon nanocomposite epoxy modified with 1-3 weight percent graphene oxide is illustrated in Figure 4.4 (a–c). The impact of machine composites on nanomaterials is demonstrated with three separate samples. The sample in Figure 4.4 (a) is a 20 μm-scale specimen that contains 1 wt.% of nanomaterial and exhibits minimal fibre debonding. Figure 4.4 (b) shows a sample with 2 weight percent nanomaterial, 20 μm scale, micro-crack, crack propagation, and machining markings. The sample in Figure 4.4 (c) has a 3-weight percent nanomaterial content and a 20 μm scale, and the machined sample fibre has suffered an insignificant fracture.

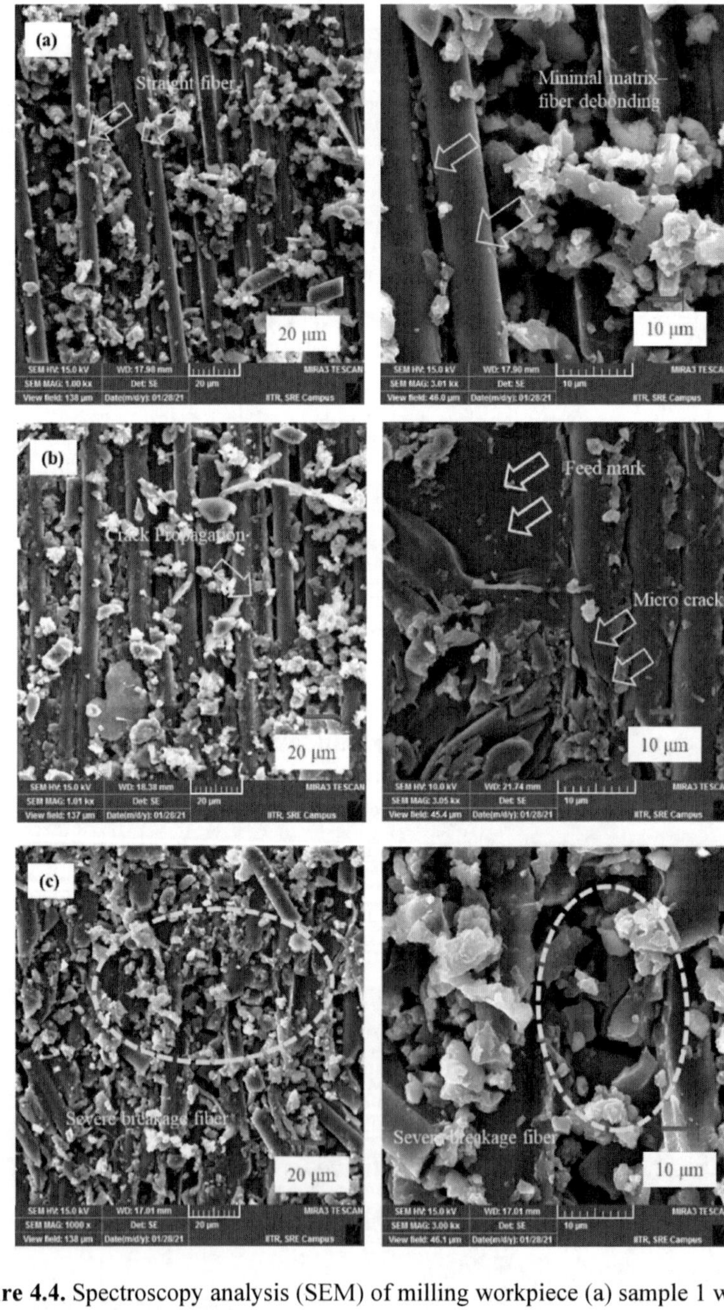

Figure 4.4. Spectroscopy analysis (SEM) of milling workpiece (a) sample 1 wt.% of GO, (b) sample 2 wt.% of GO, and (c) sample 3 wt.% of GO (J. Kumar et al., 2022; J. Kumar & Verma, 2021a).

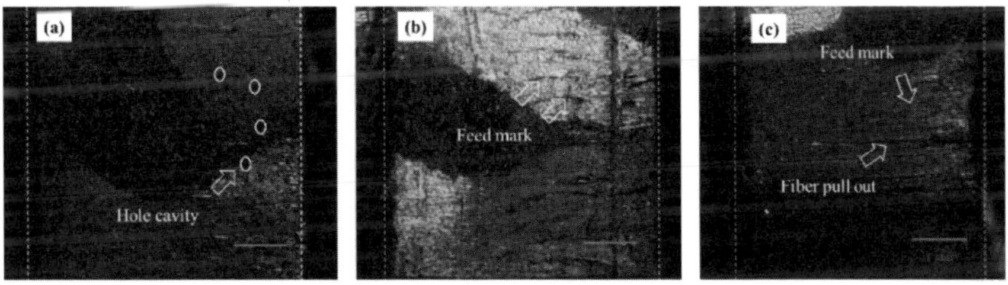

Figure 4.5. Post-milling image at lower S with wt.% of GO at (a) 1%, (b) 2%, and (c) 3% (J. Kumar et al., 2022).

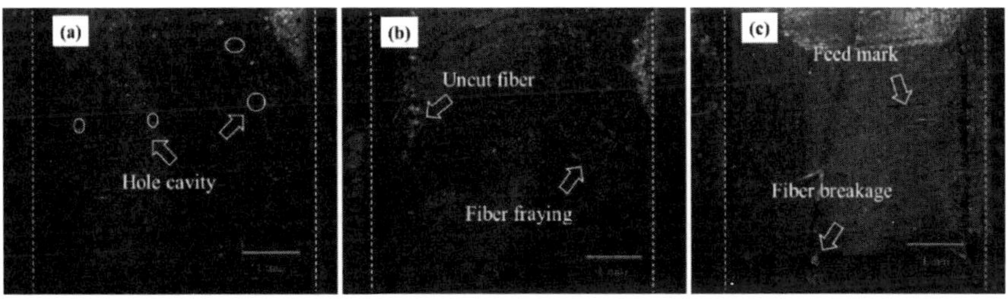

Figures 4.6. Post-milling image at medium S with wt.% of GO at (a) 1 %, (b) 2% and (c) 3% (J. Kumar et al., 2022).

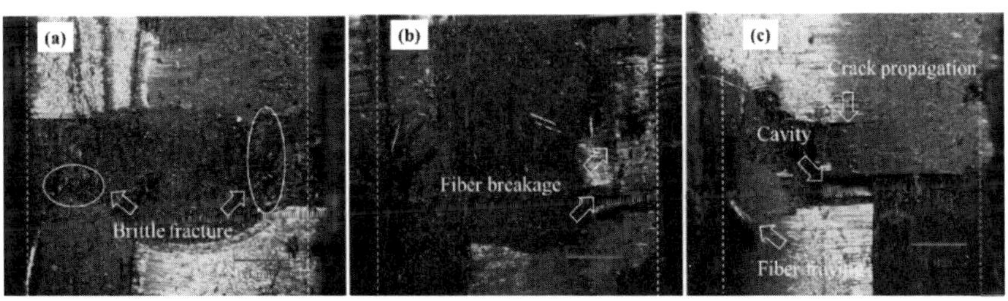

Figure 4.7. Post-milling image at higher S with wt.% of GO at (a) 1 %, (b) 2%, and (c) 3% (J. Kumar et al., 2022).

The micro-degradation of the fiber or composite interfaces at different milling operation levels was demonstrated by observing the machined sample under a microscope at varied cutting circumstances. In Figure 4.5 (a-c), we can see the common types of damage to the machine. The milling experiments showed higher surface roughness based on microscopy views of fiber pullout, uncut fiber, fraying, feed marks, cavity holes, fiber fractures, and cracks at different S. Increased F levels and S could be the cause of surface degradation. The process of creating composite materials also results in material flaws. Figure 4.6 (a-c) displays the slot surfaces as microscope images. Surface cavities, feed marks, and fiber pullout damage have all been noticed at the slot region. Increased F, inadequate fiber support, and interfacial debonding of fibers are the causes of this. For LPC, a medium S is advised to achieve the least amount of surface roughness (Palanikumar, 2012). Figure 4.7 (a-c) illustrates pullout, fiber breaking, and the development of a relatively small number of cavities. The surface degradation that increased and widened the pores in the laminates is depicted in Figure 4.7 (a-c). Unfavorably, a poor surface finish is the end consequence of cutting at a fast S, which also increases breaking, pullout, cavity formation, and brittle fracture. With rising cutting temperatures, increased F, and the aggregation of nanomaterials, the resin action causes it to become softer.

4.3. Machining and Machinability Evaluation of Fiber-Reinforced Polymer Nanocomposites

Table 4.1. Machining of LPC/LPNC

Parameters	Polymer composite		Response	Remarks	Ref.
	Nano material	Fiber			
S, F, t	-	*Glass*	T_h, T, R_a	F&S contributes 44.16% and 27.68%	(Bhat et al., 2019)
S, F, Wt.%, D	*Fly ash*	*Carbon*	T_h, T, F_d	Fly ash filled enhanced machinability	(Rajmohan, 2019)
S, F, θ	*G and GO*	*Carbon*	T_h, T, F_d	An increase in the S, F, and θ increases the T_h, T and F_d	(Çelik et al., 2019)
S, F, CNF	*CNF*	-	F_d	F_d suppressed with wt.% CNF	(Paul Theophilus Rajakumar et al., 2013)
S, F, D_c	*MWCNT*	-	F_d, T	Response value increased with S, F and D_c	(Uysal et al., 2019)
S, F	-	*Glass*	F_d	F_d is 1.8–1.9 at lower F	(Mudhukrishnan et al., 2020)
With CNT and Without CNT	*CNT*	*Glass*	T_h	T_h is lower 3% with CNT	(X. Wang et al., 2020)
Wt.%, S, F	*MWCNT*	*Carbon*	T_h, T, R_a, F_d	T_h and T decrease with the nanomaterial and Fd & SR is lower with MWCNT	(D. Kumar & Singh, 2019)
Wt.%, F, S, D	*MWCNT*	*Glass*	F_d, T_h, *FS*	F is the highest influencing factor for T_h and F_d	(Heidary et al., 2018)
S, F	*CNT*	*Carbon*	T_h, F_d	Both are controlled by controlling the F	(Kaybal et al., 2019)

Machined LPC was the primary focus of several researchers. There are many areas of engineering where polymer materials have recently been pivotal. Drilling problems with quality, accuracy, and delamination in CFRP composite material were investigated. The hybrid weighted Principal component analysis (PCA) and fuzzy inference system (FIS) methods were used to aggregate responses. MIYAKE et al., (Miyake et al., 2016) attempted machining to quantitatively investigate drilling-induced damage by evaluating residual stress details in unidirectional CFRP fibers. Through micro-Raman spectroscopy, stress distributions along the drilled-hole periphery were assessed in μm spatial resolution. Kumar and Singh (D. Kumar & Singh, 2019) used creative techniques to enhance IFSS and ILSS by combining polymer matrix multifaceted CNT. With this method, they performed an experimental assessment, and IFSS and ILSS increased by 24% and 28%, respectively. The results demonstrated that the machined hole's surface roughness improved with 1.5 wt.% of the multiwall carbon nanotube. Quan et al., (Quan & Sun, 2010) used an infiltration method with a gold chloride solution to explain the damage as delamination at the MWCNT drill hole. The delamination was investigated at the hole entrance and exit by the impacts of S and F and the tool geometry during drilling. Researchers discovered that the extent of injury increased as the hole's entry or exit approached. The experimental results suggest that the thrust is less in vibration drilling than in conventional drilling. Additionally, it was claimed that the vibration drilling technique is the most effective method for the production of micro-holes in FRPs (X. Wang et al., 2004). Palanikumar (Palanikumar, 2010) attempted to examine the Fd and surface roughness model in GFRP composite machining using the surface response methodology (RSM). The study was considered to be conducted using a central composite design with three factors and five levels. In drilling GFRP composites, the concluded F most affects the delamination variable and surface roughness. They also provided descriptions of multiple response diagrams and three-dimensional interaction plots to understand the parameters better. Ragunath et al., (Ragunath et al., 2017) suggested a streamlined optimization approach for evaluating fiber/polymer composite delamination during machining. The analysis confirms that F is the most crucial component among limitations such as drill bit diameter, point angle, and chisel edge. Kumar and Singh (Singh et al., 2017) demonstrated that the issue arose while cutting a composite material made of several carbon fibers bonded with a polymer. The delamination issue was encountered during the sample drilling. Table 4.1 provides a summary of the significant studies on the machining of LPC and nanocomposites. This table describes the considerable selection of process parameters and their impact on machinating and machinability enhancement in polymer composite/nanocomposite.

Conclusion

This chapter investigates the defects and damages occurring while machining the LPNC. The drilling and milling experiments were performed at varying process constraints to study the machining process and performance. The main emphasis is placed on the control of process parameters to reduce the defects and machining-induced damages generated in the laminated polymers. The delamination and other surface damages were explored in this section to evaluate the machinability of composites. Optimal machining indices are evaluated by analyzing the impact of process parameters, including F, S, and the weight percentage of nanomaterials. The results presented in this chapter allow us to conclude:

- The increasing F could increase the delamination damage in laminate polymers. It is mainly due to the chisel edge of the drill bit. The higher F in laminate composites causes fast contact between the chisel edge and the material surface. The material which is to be cut is extruded fastly. The cutting force pushes the tool on the workpiece, which creates high pressure on the workpiece that is to be machined. This causes the bending of the laminates and leads to crack generation in the laminates. The F considerably influences the delamination, as the punch effect significantly increases drilling delamination as the F intensity rises.
- In the initial stage of the Fd, graphene oxide has a positive impact; however, it subsequently has a negative impact. The S positively affects the delamination; it softens the matrix material. Less cutting force is also needed because the faster S creates a deformation zone and raises the temperature at the machining contact between the workpiece and the tool cutting edge. In LPC, the combined effect of S and F significantly increases delamination.
- The main functions of the F are to complete the surface and, at a higher S, to lessen vibration and the formation of defects and cracks. Because it increases bonding strength, the weight percentage of GO

also has a significant impact on the quality and surface condition of the machined sample.

- In terms of delamination, the best process parameters for milling LPNC are S level-3 (37.68 m/min), F level-1 (80 mm/min), D level-1 (0.5 mm), and GO % level-1 (1%).
- SEM analysis shows that introducing an appropriate amount (GO-1%) of nanomaterial to the modified epoxy makes the surface better and minimizes defects and cracks. On the other hand, adding more nanomaterial (GO-3%) makes the surface rougher, and more cracks, pores, and holes appear (due to the agglomeration effect).
- At lower F and higher S, a considerable degree of matrix (epoxy) degradation is seen. Moreover, chip breakup along the milling slot (machined sample) edge was noted at lower S and greater F values.
- A major factor in decreasing milling-induced damage is the inclusion of graphene oxide. The suggested nanocomposite can be machined more effectively with a small amount of graphene added.

The next chapter involves the appraisal of each machining parameter, using parametric optimization and predictive modeling to build a strong optimization module. It can manage multiple responses and varying constraints at the same time. This section introduces the novel optimization module responsible for the development of an efficient machining environment. This criterion shall control the damage and defects while machining polymers through appropriate conditions. The proposed module will overcome the limitations of the existing traditional optimization approach. Therefore, an endeavors is being undertaken to ensure a proficient machining (Drilling & Milling) environment to address the intricate real-world challenges that arise while machining LPNC.

References

Adak, N. C., Chhetri, S., Sabarad, S., Roy, H., Murmu, N. C., Samanta, P., & Kuila, T. (2019). Direct observation of micro delamination in graphene oxide incorporated carbon fiber/epoxy composite via in-situ tensile test. *Composites Science and Technology*, *177*, 57–65.

Altin Karataş, M., & Gökkaya, H. (2018). A review on machinability of carbon fiber reinforced polymer (CFRP) and glass fiber reinforced polymer (GFRP) composite materials. *Defence Technology*, *14*(4), 318–326.

Anarghya, A., Harshith, D. N., Rao, N., Nayak, N. S., Gurumurthy, B. M., Abhishek, V. N., & Patil, I. G. S. (2018). Thrust and torque force analysis in the drilling of aramid fibre-reinforced composite laminates using RSM and MLPNN-GA. *Heliyon*, *4*(7), 1–39.

Azmi, A. I., Lin, R. J. T., & Bhattacharyya, D. (2013). Machinability study of glass fibre-reinforced polymer composites during end milling. *International Journal of Advanced Manufacturing Technology*, *64*(1–4), 247–261.

Bhat, R., Mohan, N., Sharma, S., Agarwal, R. A., Rathi, A., & Subudhi, K. A. (2019). Multi-response optimization of the thrust force, torque and surface roughness in drilling of glass fiber reinforced polyester composite using GRA-RSM. *Materials Today: Proceedings*, *19*(2), 333–338.

Bhatnagar, N., Singh, I., & Nayak, D. (2004). Damage investigation in drilling of glass fiber reinforced plastic composite laminates. *Materials and Manufacturing Processes*, *19*(6), 995–1007.

Bielawski, R., Kowalik, M., Suprynowicz, K., Rzadkowski, W., & Pyrzanowski, P. (2017). Experimental study on the riveted joints in glass fibre reinforced plastics (GFRP). *Archive of Mechanical Engineering*, *64*(3), 301–313.

Çelik, Y. H., Kilickap, E., & Koçyiğit, N. (2019). Evaluation of drilling performances of nanocomposites reinforced with graphene and graphene oxide. *International Journal of Advanced Manufacturing Technology*, *100*(9–12), 2371–2385.

Dahnel, A. N., Ascroft, H., & Barnes, S. (2016). The Effect of Varying Cutting Speeds on Tool Wear during Conventional and Ultrasonic Assisted Drilling (UAD) of Carbon Fibre Composite (CFC) and Titanium Alloy Stacks. *Procedia CIRP*, *46*, 420–423.

Davim, J. P., & Reis, P. (2005). Damage and dimensional precision on milling carbon fiber-reinforced plastics using design experiments. *Journal of Materials Processing Technology*, *160*(2), 160–167.

Dixit, S., Mahata, A., Mahapatra, D. R., Kailas, S. V., & Chattopadhyay, K. (2018). Multi-layer graphene reinforced aluminum – Manufacturing of high strength composite by friction stir alloying. *Composites Part B: Engineering*, *136*, 63–71.

El-Sonbaty, I., Khashaba, U. A., & Machaly, T. (2004). Factors affecting the machinability of GFR/epoxy composites. *Composite Structures*, *63*(3–4), 329–338.

Eneyew, E. D., & Ramulu, M. (2014). Experimental study of surface quality and damage when drilling unidirectional CFRP composites. *Journal of Materials Research and Technology*, *3*(4), 354–362.

Erkan, Ö., Işık, B., Çiçek, A., & Kara, F. (2013). Prediction of damage factor in end milling of glass fibre reinforced plastic composites using artificial neural network. *Applied Composite Materials*, *20*(4), 517–536.

Gara, S., M'hamed, S., & Tsoumarev, O. (2018). Temperature measurement and machining damage in slotting of multidirectional CFRP laminate. *Machining Science and Technology*, *22*(2), 320–337.

Gaugel, S., Sripathy, P., Haeger, A., Meinhard, D., Bernthaler, T., Lissek, F., Kaufeld, M., Knoblauch, V., & Schneider, G. (2016). A comparative study on tool wear and laminate damage in drilling of carbon-fiber reinforced polymers (CFRP). *Composite Structures*, *155*, 173–183.

Gemi, L., Morkavuk, S., Köklü, U., & Gemi, D. S. (2019). An experimental study on the effects of various drill types on drilling performance of GFRP composite pipes and damage formation. *Composites Part B: Engineering*, *172*, 186–194.

Geng, D., Liu, Y., Shao, Z., Lu, Z., Cai, J., Li, X., Jiang, X., & Zhang, D. (2019). Delamination formation, evaluation and suppression during drilling of composite laminates: A review. *Composite Structures*, *216*, 168–186.

Ha, S. J., Kim, K. B., Yang, J. K., & Cho, M. W. (2017). Influence of cutting temperature on carbon fiber-reinforced plastic composites in high-speed machining. *Journal of Mechanical Science and Technology*, *31*(4), 1861–1867.

Haddad, M., Zitoune, R., Bougherara, H., Eyma, F., & Castanié, B. (2014). Study of trimming damages of CFRP structures in function of the machining processes and their impact on the mechanical behavior. *Composites Part B: Engineering*, *57*, 136–143.

Haeger, A., Meinhard, D., Lissek, F., Kaufeld, M., Hoffmann, M. J., Schneider, G., & Knoblauch, V. (2016). Interaction between laminate quality, drilling-induced delamination and mechanical properties in machining of carbon fibre reinforced plastic (CFRP). *Materialwissenschaft Und Werkstofftechnik*, *47*(11), 997–1014.

Haeger, A., Schoen, G., Lissek, F., Meinhard, D., Kaufeld, M., Schneider, G., Schuhmacher, S., & Knoblauch, V. (2016). Non-destructive detection of drilling-induced delamination in CFRP and its effect on mechanical properties. *Procedia Engineering*, *149*, 130–142.

He, Y., Qing, H., Zhang, S., Wang, D., & Zhu, S. (2017). The cutting force and defect analysis in milling of carbon fiber-reinforced polymer (CFRP) composite. *International Journal of Advanced Manufacturing Technology*, *93*(5–8), 1829–1842.

Heidary, H., Zarif, N., Minak, G., Karimi, N. Z., & Minak, G. (2018). Investigation on delamination and flexural properties in drilling of carbon nanotube/polymer composites. *Composite Structures*, *201*, 112–120.

Ho-Cheng, H., & Dharan, C. K. H. (1990). Delamination during drilling in composite laminates. *Journal of Manufacturing Science and Engineering, Transactions of the ASME*, *112*(3), 236–239.

Hocheng, H., & Tsao, C. C. (2003). Comprehensive analysis of delamination in drilling of composite materials with various drill bits. *Journal of Materials Processing Technology*, *140*, 335–339.

Isbilir, O., & Ghassemieh, E. (2012). Delamination and wear in drilling of carbon-fiber reinforced plastic composites using multilayer TiAlN/TiN PVD-coated tungsten carbide tools. *Journal of Reinforced Plastics and Composites*, *31*(10), 717–727.

John, K. M., Kumaran, S. T., Kurniawan, R., Moon Park, K., & Byeon, J. H. (2019). Review on the methodologies adopted to minimize the material damages in drilling of carbon fiber reinforced plastic composites. *Journal of Reinforced Plastics and Composites*, *38*(8), 351–368.

Kaybal, H. B., Ünüvar, A., Koyunbakan, M., & Avcı, A. (2019). A novelty optimization approach for drilling of CFRP nanocomposite laminates. *International Journal of Advanced Manufacturing Technology*, *100*(9–12), 2995–3012.

Khashaba, U. A. (2004). Delamination in drilling GFR-thermoset composites. *Composite Structures*, *63*(3–4), 313–327.

Kostagiannakopoulou, C., Tsilimigkra, X., Sotiriadis, G., & Kostopoulos, V. (2017). Synergy effect of carbon nano-fillers on the fracture toughness of structural composites. *Composites Part B: Engineering*, *129*, 18–25.

Krishnamoorthy, A., Rajendra Boopathy, S., Palanikumar, K., & Paulo Davim, J. (2012). Application of grey fuzzy logic for the optimization of drilling parameters for CFRP composites with multiple performance characteristics. *Measurement: Journal of the International Measurement Confederation*, *45*(5), 1286–1296.

Kumar, D., & Singh, K. K. (2019). Investigation of delamination and surface quality of machined holes in drilling of multiwalled carbon nanotube doped epoxy/carbon fiber reinforced polymer nanocomposite. *Proceedings of the Institution of Mechanical Engineers, Part L: Journal of Materials: Design and Applications*, *233*(4), 647–663.

Kumar, J., Abhishek, K., Xu, J., & Verma, R. K. (2022). Experimental Investigation on Machine-Induced Damages during the Milling Test of Graphene/Carbon Incorporated Thermoset Polymer Nanocomposites. *Journal of Composites Science*, *6*(3), 1–12. https://doi.org/10.3390/jcs6030077.

Kumar, J., & Verma, R. K. (2021a). Delamination assessment during machining of laminated polymer nanocomposite. *International Journal of Modern Manufacturing Technologies*, *13*(2), 109–115. https://doi.org/10.54684/ijmmt.2021.13.2.109.

Kumar, J., & Verma, R. K. (2021b). Experimental investigation for machinability aspects of graphene oxide/carbon fiber reinforced polymer nanocomposites and predictive modeling using hybrid approach. *Defence Technology*, *17*(5), 1671–1686. https://doi.org/10.1016/j.dt.2020.09.009.

Liu, D. F., Tang, Y. J., & Cong, W. L. (2012). A review of mechanical drilling for composite laminates. *Composite Structures*, *94*(4), 1265–1279. https://doi.org/10.1016/j.compstruct.2011.11.024.

Meinhard, D., Haeger, A., & Knoblauch, V. (2021). Drilling induced defects on carbon fiber-reinforced thermoplastic polyamide and their effect on mechanical properties. *Composite Structures*, *256*, 113138.

Miyake, T., Mukae, K., & Futamura, M. (2016). Evaluation of machining damage around drilled holes in a CFRP by fiber residual stresses measured using micro-Raman spectroscopy. *Mechanical Engineering Journal*, *3*(6), 16–00301.

Mudhukrishnan, M., Hariharan, P., & Palanikumar, K. (2020). Measurement and analysis of thrust force and delamination in drilling glass fiber reinforced polypropylene composites using different drills. *Measurement: Journal of the International Measurement Confederation*, *149*, 106973.

Palanikumar, K. (2006). Cutting parameters optimization for surface roughness in machining of GFRP composites using Taguchi's method. *Journal of Reinforced Plastics and Composites*, *25*(16), 1739–1751.

Palanikumar, K. (2010). Modeling and analysis of delamination factor and surface roughness in drilling GFRP composites. *Materials and Manufacturing Processes*, *25*(10), 1059–1067.

Palanikumar, K. (2011). Experimental investigation and optimisation in drilling of GFRP composites. *Measurement: Journal of the International Measurement Confederation*, *44*(10), 2138–2148.

Palanikumar, K. (2012). Analyzing surface quality in machined composites. In *Machining Technology for Composite Materials*. Woodhead Publishing Limited.

Papadopoulos, A., Gkikas, G., Paipetis, A. S., & Barkoula, N. M. (2016). Effect of CNTs addition on the erosive wear response of epoxy resin and carbon fibre composites. *Composites Part A: Applied Science and Manufacturing*, *84*, 299–307.

Paul Theophilus Rajakumar, I., Hariharan, P., & Srikanth, I. (2013). A study on monitoring the drilling of polymeric nanocomposite laminates using acoustic emission. *Journal of Composite Materials*, *47*(14), 1773–1784.

Puw, H. Y., & Hocheng, H. (1993). Machinability Test Of Carbon Fiber-Reinforced Plastics In Milling. *Materials and Manufacturing Processes*, *8*(6), 717–729.

Quan, Y., & Sun, L. (2010). Investigation on drilling-induced delamination of CFRP with infiltration method. *Advanced Materials Research*, *139–141*, 55–58.

Ragunath, S., Velmurugan, C., & Kannan, T. (2017). Optimization of drilling delamination behavior of GFRP/clay nano-composites using RSM and GRA methods. *Fibers and Polymers*, *18*(12), 2400–2409.

Rajesh Mathivanan, N., Mahesh, B. S., & Anup Shetty, H. (2016). An experimental investigation on the process parameters influencing machining forces during milling of carbon and glass fiber laminates. *Measurement: Journal of the International Measurement Confederation*, *91*, 39–45.

Rajmohan, T. (2019). Experimental investigation and optimization of machining parameters in drilling of fly ash-filled carbon fiber reinforced composites. *Particulate Science and Technology*, *37*(1), 21–30.

Ray, D., Puri, A. B., & Nagahanumaiah. (2019). Investigation on Cutting Forces and Surface Finish in Mechanical Micro Milling of Zr-Based Bulk Metallic Glass. *Journal of Advanced Manufacturing Systems*, *18*(1), 113–132.

Razfar, M. R., & Zadeh, M. R. Z. (2009). Optimum damage and surface roughness prediction in end milling glass fibre-reinforced plastics, using neural network and genetic algorithm. *Proceedings of the Institution of Mechanical Engineers, Part B: Journal of Engineering Manufacture*, *223*(6), 653–664.

Santhanakrishnan Balakrishnan, V., Seidlitz, H., Yellur, M. R., & Vogt, N. (2019). A study on the influence of drilling and CO 2 laser cutting in carbon/epoxy laminates. *Journal of Materials Research and Technology*, *8*(1), 944–949.

Saoudi, J., Zitoune, R., Mezlini, S., Gururaja, S., & Seitier, P. (2016). Critical thrust force predictions during drilling: Analytical modeling and X-ray tomography quantification. *Composite Structures*, *153*, 886–894.

Saravanakumar, K., Subramanian, H., Arumugam, V., & Dhakal, H. N. (2019). Influence of milled glass fillers on the impact and compression after impact behavior of glass/epoxy composite laminates. *Polymer Testing*, *75*, 133–141.

Sawangphruk, M., Suksomboon, M., Kongsupornsak, K., Khuntilo, J., Srimuk, P., Sanguansak, Y., Klunbud, P., Suktha, P., & Chiochan, P. (2013). High-performance supercapacitors based on silver nanoparticle-polyaniline- graphene nanocomposites coated on flexible carbon fiber paper. *Journal of Materials Chemistry A*, *1*(34), 9630–9636.

Sheikh-Ahmad, J. Y., Dhuttargaon, M., & Cheraghi, H. (2017). New tool life criterion for delamination free milling of CFRP. *International Journal of Advanced Manufacturing Technology*, *92*(5–8), 2131–2143.

Singh, V., Bhandari, R., & Yadav, V. K. (2017). An experimental investigation on machining parameters of AISI D2 steel using WEDM. *International Journal of Advanced Manufacturing Technology*, *93*(1–4), 203–214.

Sorrentino, L., Turchetta, S., & Bellini, C. (2018). A new method to reduce delaminations during drilling of FRP laminates by feed rate control. *Composite Structures*, *186*, 154–164.

Stone, R., & Krishnamurthy, K. (1996). A neural network thrust force controller to minimize delamination during drilling of graphite-epoxy laminates. *International Journal of Machine Tools and Manufacture*, *36*(9), 985–1003.

Thakur, R. K., Singh, K. K., & Kumar, K. (2020). Investigation of milling characteristics in graphene-embedded epoxy/carbon fibre reinforced composite. *Materials Today: Proceedings*, *33*, 5643–5648.

Uysal, A., Eryildiz, E., & Altan, E. (2019). Turning Performance of Bonded Cutting Tools with Nanographene or Multi-walled Carbon Nanotube Particle-Reinforced Epoxy-Based Nanocomposite Adhesives. *Arabian Journal for Science and Engineering*, *44*(9), 7737–7752.

Voß, R., Henerichs, M., Kuster, F., & Wegener, K. (2014). Chip root analysis after machining carbon fiber reinforced plastics (CFRP) at different fiber orientations. *Procedia CIRP*, *14*, 217–222.

Wang, F. ji, Zhang, B. yu, Zhao, M., Cheng, D., & Wang, Z. gang. (2019). Evolution laws of fiber-matrix interface cracks in machining of carbon fiber reinforced polymer. *International Journal of Advanced Manufacturing Technology*, *101*(1–4), 963–977.

Wang, X., Melly, S. K., Li, N., Wang, G. D., Peng, T., Li, Y., & Zhao, Q. Di. (2020). Helical milling response of glass fiber-reinforced polymer composite with carbon nanotube buckypaper interlayer. *Polymers and Polymer Composites*, *28*(6), 378–387.

Wang, X., Wang, L. J., & Tao, J. P. (2004). Investigation on thrust in vibration drilling of fiber-reinforced plastics. *Journal of Materials Processing Technology*, *148*(2), 239–244.

Xu, J., & El Mansori, M. (2016). Numerical studies of frictional responses when cutting hybrid CFRP/Ti composite. *International Journal of Advanced Manufacturing Technology*, *87*(1–4), 657–675.

Zhang, K., Gu, Y., Li, M., Wang, S., & Zhang, Z. (2019). Effects of curing time and de-molding temperature on the deformation of glass fiber/epoxy resin prepreg laminates fabricated by rapid hot press. *Polymers and Polymer Composites*, *27*(6), 301–313.

Chapter 5

Development of Robust Optimization Modules for Parametric Appraisal and Multi-Objective Optimization: Mathematical Modeling

5.1. Introduction to Parametric Optimization

In the rapidly expanding manufacturing sectors of the present day, numerous polymer industries are implementing lean, agile, and flexible production environments to adapt to their consumers' rapidly evolving demands and requirements. Customer satisfaction is crucial for any organization or enterprise regarding procedures or goods. The utilization of an automated, computerized, controlled production system is necessary to meet the rapidly evolving demands of consumers (Mourtzis et al., 2014). The machining cost, operator experience, and setup/installation cost of these precision machinery types are relatively high. It is difficult for the manufacturer to calculate the payback cost from these configurations. Selection of optimal process parameters and integration of conflicting machining responses depend on the domain of experiments, empirical expressions to investigate accurate constraints, optimization tool selection, and statistical analysis. The current methods have specific limitations and drawbacks, including the assignment of weights to responses, which is typically contingent upon the decision-making process (DM), and the unfeasibility of negligible response correlation in real-world scenarios (Datta et al., 2008).

In optimizing different process constraints, eminent scholars made many attempts to evaluate the optimal conditions for an efficient manufacturing environment. Predictive modeling of machining operations is the first and most important step towards process control and optimization. An output response and the input process parameters are correlated in a predictive model. Quantitatively predicting the output of machining operations should be the primary goal of developing models. Modeling could be helpful for optimal efficiency, quality, and cost-effectiveness in the machining process planning. Because machining process modeling is lacking, the underlying mechanisms of tool tip and work material interactions remain unclear (Altintas et al., 1998). For practical applications, the predictive modeling of machining operations involves two steps. Models for machining output are established after developing predictive models for machining parameters. During the first stage of the process, several phenomena such as temperature changes, friction, tool chip contact time, chip flow, and tensions are anticipated to occur. Stage 2 involves predicting machining performance parameters, such as surface roughness, chip form/chapping, cutting forces, torque, strength, and tool wear/life, that are required for practical application. The workpiece's dimensions, shape, and qualities significantly impact how well it machines. The machining tool's characteristics can be influenced by material, design, geometry, and lubricants. The links between process inputs and outputs can be established using a variety of approaches. The highly non-linear characteristic of metal-cutting processes means that there aren't many studies that can reliably carry out the function that was specified above. Analytical, numerical, and empirical modeling are the three primary steps used to represent shifting constraints. According to Luttervelt et al., (Altintas et al., 1998), with 43% engaged in empirical modeling, 32% in analytical modeling, and 18% in numerical modeling, research organizations were quite active in this area.

The variables of interest in this study are the drill diameter, feed rate, and drilling velocity. A hybrid methodology combining the Weighted principal component analysis (WPCA) and the fuzzy inference system (FIS) method was utilized to aggregate various products and quality attributes. To determine the machining reaction behavior, the fuzzy idea has successfully investigated numerous expert norms and opinions (Abhishek et al., 2016). The environmental strategy and various input parameters were tested. An optimization method has been implemented to evaluate the machining factors, including voltage, wire input, and welding velocity. They described the Taguchi method for removing correlated reactions using the weightage principal component analysis (WPCA) module. A confirmatory test was used to verify the machining process results, which was carried out using the L_{16} orthogonal array (Routara et al., 2010). Plasma arc cutting's machining constraints were optimized by implementing Taguchi concepts. The WPCA-based Taguchi approach was effectively employed to convert multiple responses into a single objective function, and the L27 orthogonal array was used to conduct experiments (M. K. Das et al., 2014). A response correlation study used PCA-coupled GRA techniques to ascertain the optimal parametric combination (Kharwar & Verma, 2020). The material structure behaviors are readily described by the electric discharge machining (EDM) process, which is investigated with Taguchi's strategy to optimize WPCA techniques (D. Lahane, 2012). A hybrid optimization tool was suggested to optimize the machining constraint during the turning process (Vasudevan et al., 2018). The mechanical and tribological properties of graphite and carbon composites were successfully implemented (Yang et al., 2010).

Various robust optimization modules, including ANN, fuzzy, GA, SA, and PSO algorithms, are developed by the researchers in response to the study of tool wear, fiber draw out, matrix debonding, and other variables (Kharwar & Verma, 2019). Researchers have used utility theory, desirability function analysis (DFA), and the method for order of preference by similarity to ideal solution (TOPSIS) to look into how input, tool components, and cutting speed affect the way machining responds. Most of the research has focused on individual optimization modules; consequently, information is scarce regarding hybridization strategies (Weglowski, 2014). It is believed that the machining characteristics can be significantly improved by implementing appropriate parametric parameters using hybridization modules. Several pioneering academics attempted to conduct experimental investigations on the machining of polymeric composites (Kharwar & Verma, 2019). The desired machining properties are achieved, such as a higher material removal rate, torque, surface roughness, and a lower thrust fore value. By measuring the rate of MRR and SR, the productivity and quality of the manufactured component are determined. Among other aesthetic qualities, precise fits, tolerance, rivet and role fastening, and drill hole finishing are all regarded as critical design elements for a unique item that will be subjected to creep and fatigue (Bielawski et al., 2017).

5.2. Predictive Modeling

In the past twenty years, scientists have developed various techniques to address the intricate optimization problems in power, control, robotics, engineering, medicine, etc. (Papageorgiou et al., 2019). Various conventional mathematical methods were employed in the early research on machining and machinability; however, they are time-consuming for resolving complex issues (Kulkarni et al., 2020). Many branches of engineering have adopted and used new, revolutionary methods in the last few decades. Consequently, hybrid optimization can exhibit unparalleled reliability and efficacy by implementing engineering optimization techniques. To enhance the effectiveness of optimization techniques, pioneer scholars meticulously define hybrid modules. Adaptive optimization or improvement techniques may be implemented to identify optimal parameters for improved machining efficiency. Tool life and production costs can be effectively balanced by selecting process parameters.

Graphene oxide (GO) is possible. It has a lot of potential uses in engineering fields, including biomedical, sensors, batteries, structure applications in cars and planes, electromagnetic (EM) protection, and fuselage components (Jin et al., 2021; Khan et al., 2019). There is no comprehensive data regarding incorporating MWCNT/CNO/polymer nanocomposites. According to published research, renowned academics made significant attempts to machine FRP by taking into account competing machining performance factors, including MRR, SR, cutting force, circularity error, etc., using a variety of optimization tools like DFA, GRA, TOPSIS, Utility, and Taguchi (Datta & Sankar Mahapatra, 2010; Jenarthanan & Jeyapaul, 2018; Naveen Sait et al., 2009; Rajkumar et al., 2017; Singaravel & Selvaraj, 2016). However, there is a lack of research on the machining features of MWCNT & CNO composites and the concurrent optimization of these innovative materials' machining properties. Furthermore, it has been determined that several studies operate under the assumption of a uniform priority weight assignment and negligible response correlation. During multiple machining performance optimizations of contradictory responses. This is not likely in real practice and results in ambiguity, inaccuracy, and solution errors. Nevertheless, strong optimization modules still do not manage the competing replies' priority weight problems. This study's investigation of the PCA technique has effectively addressed these critical difficulties. The machinability features go through a first phase, and the remarkable qualities of these new nanocomposites are not adequately investigated in terms of utilization. As a result, the polymer industry, academics, and working engineers may find it interesting to conduct research, development, and commercialization in this field. The two most dominant machining processes for polymer composites, i.e., drilling and milling, are attempted in the present work. The machinability and machining aspect-related study of proposed polymer nanocomposites has a vast scope of research. Heidarya et al., (Heidary et al., 2018) conducted tests in drilling woven E-glass epoxy fiber composites with multi-wall nanotubes for analyzed thrust, delamination, and residual flexural strength. The process parameters for the experiment were the feed rate, the speed of the spindle, the drill diameter, and the weight % of the carbon nanotubes. A three-point bending test was used to determine the flexural strength of the drilled samples in experiments conducted using the Taguchi approach. Finally, Grey's theory was used to control the varying constraints for desired machining characteristics during drilling tests. They concluded that the feed rate has the highest effect on the thrust force and delamination factor, trailed by spindle speed. Rajkumar et al., (Rajkumar et al., 2017) considerably improved cutting process parameters for CFRP composite material, including MRR and surface roughness, by utilizing a desirability module based on Taguchi. The CNC machine was used for the experiment since micro-drilling is required. Desirability-based functional analysis is instrumental in optimizing multiple response problems. As input process parameters, this machining operation considers cutting speed, depth of cut, and feed rate. In the end, they used analysis of variance (ANOVA) to examine the correlation between process parameters and performance traits. They found that feed rate, spindle speed, and air pressure affected the drilling parameter by 39.21%, 49.32%, and 8.58%, respectively. Table 5.1 provides an overview of the many modules famous scholars use to optimize process parameters in machining polymer composites and nanocomposites. This table describes the selection of optimization modules and their significance in machining and machinability enhancement in polymers and nanocomposites.

A high strength-to-weight ratio, high aspect surface area, and mechanical properties are required for multifunctional applications. Eminent scholars have researched the development characterization of macro-size fiber and conventional composite materials during the last two decades. However, limited research exists on manufacturing polymer nanocomposites, their machining and machinability aspects, and parametric control through optimization modules. Furthermore, conventional optimization approaches were used to optimize machining processes. The cutting tool material is coated, and the workpiece is a hybrid composite. Laminated polymer nanocomposite is one of the most recently studied products in machining, but there are few applications in the industry due to its lack of exploration. Critical parameters such as cutting speed, feed rate, tool diameter, tool materials coating, etc., have always been highly required to focus the process models; however, to be beneficial to the industry, these variables must be linked to performance measures (MRR, surface roughness, cutting power, delamination defects) and product quality. Due to inadequate machining information, complexity of factors, and conflicting response behavior, machining performance is altered. High production costs, poor product quality, and energy inefficiency could result from inadequate knowledge. These problems pushed for analyzing and optimizing the machining process's performance to maximize profit and improve product quality. This section is designed to optimize the process constraints during the machining of hybrid laminated polymer nanocomposite.

Table 5.1. Optimization of process parameters through various modules in polymeric material machining

Module	Operation	Composite material	Machining		Ref.
			Parameters	Response	
GRA-Fuzzy	Turning	GFRP/epoxy	S, Nr, F, D_c	R_a, T_h, MRR	(Vasudevan et al., 2014)
TOPSIS			S, F, D_c	R_a, MRR	(A. K. Parida & Routara, 2014)
Taguchi approach		Modified epoxy with fly ash/GFRP		R_a	(A. Kumar Parida et al., 2014)
TOPSIS Deng similarity	Drilling	GFRP/epoxy	S, F, t, D	R_a, T_h, T, F_d	(Sonkar et al., 2014)
Taguchi-GRA			S, F	R_a, T_h, F_d	(Palanikumar, 2011)
Taguchi approach			S, F, θ	F_d	(Kilickap, 2010)
			S, F, θ, w	T_h, T, R_a, C_e	(Vankanti & Ganta, 2014)
			t, S, F, D	T_h, T	(Mohan et al., 2005)
Fuzzy logic			S, F, D	T_h	(Latha & Senthilkumar, 2009)
Taguchi SQP algorithm		CFRP/epoxy	S, F, G_t	F_d, R_a, T_h	(Enemuoh et al., 2001)
Taguchi-GA			S, F	F_d (I_n and Out), C_e, H_d	(Krishnaraj et al., 2012)
GA			S, F	MRR, F_d	(Sardiñas et al., 2006)
GRA-Fuzzy			S, F, θ	T_h, F_d (In and Out), e	(Krishnamoorthy et al., 2012)
Taguchi-ANN			S, F, D	T_h, R_a	(Tsao & Hocheng, 2008)
Taguchi approach			S, F, θ	F_d	(Gaitonde et al., 2008)
	Milling	MWCNT/epoxy	S, F, D_c	R_a	(Harii Krishna Rao & Ansari, 2013)
		GFRP/epoxy	S, F, D_c	R_a, TL, C_f	(Azmi et al., 2013)
Taguchi-GRA		GO/CFRP/epoxy	S, F, wt.% GO	F_d, R_a	(Thakur et al., 2020)

The theories of delamination and other machining-induced damages were thoroughly examined by a variety of distinguished scholars (Xu et al., 2018, 2019a, 2020). Nevertheless, the delamination continues to cause the manufacturing industries to be overly concerned, directly impacting productivity and quality. Mechanical behavior is necessary to increase machining efficiency (drilling) and overcome this new polymer nanocomposite's damage/defect and waste. Only through experiments and the objectives of the statistical model between process constraints and response is it possible (Sorrentino et al., 2017). In addition, the literature in this field is inadequate and requires further attention. Consequently, examining this intricate matter in nanocomposites is imperative by employing efficient optimization techniques and instruments. Developing an efficient machining model and an optimal set of process parameters to regulate the impact of damage is highly desirable to gain a more comprehensive understanding of these novel material composites. It is insufficient to experiment with process parameters to assess their machinability. Consequently, this section introduces a predictive mathematical model to boost process efficiency.

The fabrication of composite materials has been the subject of research by various researchers, who have examined the mixing of multiple methods and the proportion of varying weight percentages of graphene in the CF matrix to enhance the mechanical, electrical, and surface tropology properties. Several researchers have illustrated their work at international and national levels. Nevertheless, a comprehensive literature review has revealed that only a handful of studies have examined the potential of graphene oxide nanofillers in glass/carbon fiber composites. According to an extensive literature assessment, pioneer researchers synthesized and developed nanomaterial graphene oxide. Graphene oxide can revolutionize several technical fields, including biomedicine, sensors, batteries, and structural applications in automobiles and airplanes. Lightweight, strong, and long-lasting CFRP materials are a great alternative to heavy metals and their alloys because of these qualities. Nanocomposites made of graphene oxide and carbon fiber polymer have not been extensively studied. Renowned researchers have devoted a great deal of time and energy to optimizing CFRP machining by considering competing machining performance metrics (e.g., MRR, SR, cutting force, tool wear, etc.) and several optimization methods (e.g., DFA, GRA, TOPSIS, Utility, Taguchi). However, there is not enough literature on machining graphene-modified epoxy/CFRP composites and the simultaneous optimization of these unique materials' machining properties. Additionally, when optimizing the performance of contradictory responses through multiple machining, it has been discovered that most studies make unrealistic assumptions of zero response correlation and uniform priority weight assignment. This leads to solutions that are ambiguous, inaccurate, and erroneous. However, strong optimization modules that can deal with the priority weight problems caused by the competing solutions are still in low supply. To fully use the remarkable features of these new nanocomposites, the machinability issues go through an initial stage, and the work is not thoroughly investigated. Consequently, it has the makings of a promising R&D and commercialization focus for engineers in the field, the polymer industry, and academic institutions.

5.2.1. Mathematical Modeling of Drilling Parameters

The impact of varying process parameters on drilling performances was investigated following the drilling of polymer nanocomposites. The influence factor on machining performance was examined using an ANOVA test with a 95% confidence level (Ajith Arul Daniel et al., 2019). The F-value (Fisher test) was implemented to evaluate the maximal value of the machining parameters, which indicates the most significant factors that influence the response (Liu et al., 2010). However, the P-value suggests that the process parameter is statistically significant at less than 0.05. Table 5.2 discusses the thrust (Th) and torque (T) variance analyses. The two most important factors influencing thrust force are feed rate (30.06%) and spindle speed (46.71%). The feed rate (84.54%) and spindle speed (8.85%) have the most significant effects on torque in Table 5.2, with graphene oxide weight percentage (3.58%) being the second most crucial factor. Table 5.2 shows that the delamination factor at the exit is 11.00%, while at the entrance, it is 51.08%, spindle speed is 26.69%, and weight percent graphene oxide is 8.39%. Table 5.2 shows that the same order of variables affects the delamination factor at the exit, while at the entrance, it is 51.08%, spindle speed is 20.65%, and weight percent graphene oxide is 11.64%. From Table 5.2, the surface roughness (R_a) is primarily affected by the feed rate (65.89 %), wt. % of graphene oxide is (18.43 %) followed by spindle speed (12.64 %) and ten-point scale surface roughness (R_z) primarily affected by feed rate (50.23 %), spindle speed (40.46 %) followed by wt. % of graphene oxide is 3.24 %, respectively. In Table 5.2, circularity error is primarily affected by spindle speed (67.17 %), feed rate (30.51 %), and trail by wt.% of graphene oxide (0.81%). The attribute values of the suggested (Th, T, Fd (IN), Fd (OUT), Ra, Rz, and Ce) models are quite near to 1, making them highly acceptable. It is highly desirable to be able to examine the R^2 value as a linear model. The fact that the P-value is less than 0.05 suggests that the mean response of the machining output is affected by one or more model parameters, which is supported by the data. Additional exploration has been justified by satisfactory model adequacy. Further analysis can be conducted using the results gained from the variance analysis in this study to determine the ideal setting. While drilling GO/CF polymer nanocomposites, it can improve hole quality by lowering thrust, torque, delamination factor (IN & OUT), machined roughness (Ra & Rz), and circularity error (Ce).

Table 5.2. ANOVA for CNC drilling process (J. Kumar & Verma, 2021a)

Source	DF	Seq SS	Contribution	Adj SS	Adj MS	F-Value	P-Value
Thrust force							
Model	3	2364.95	86.86%	2364.95	788.32	24.25	0.000
Linear	3	2364.95	86.86%	2364.95	788.32	24.25	0.000
S	1	1271.84	46.71%	1271.84	1271.84	39.12	0.000
F	1	818.51	30.06%	818.51	818.51	25.17	0.000
GO	1	274.60	10.09%	274.60	274.60	8.45	0.014
Error	11	357.64	13.14%	357.64	32.51	6.83	0.134
Total	14	2722.59	100.00%				
Torque							
Model	3	0.016575	96.97%	0.016575	0.005525	117.25	0.000
Linear	3	0.016575	96.97%	0.016575	0.005525	117.25	0.000
S	1	0.001513	8.85%	0.001512	0.001512	32.10	0.000
F	1	0.014450	84.54%	0.014450	0.014450	306.66	0.000
GO	1	0.000612	3.58%	0.000612	0.000612	13.00	0.004
Error	11	0.000518	3.03%	0.000518	0.000047	0.35	0.889
Total	14	0.017093	100.00%				
Delamination factor (IN)							
Model	3	0.006194	86.36%	0.006194	0.002065	23.22	0.000
Linear	3	0.006194	86.36%	0.006194	0.002065	23.22	0.000
S	1	0.001928	26.89%	0.001928	0.001928	21.69	0.001
F	1	0.003664	51.08%	0.003664	0.003664	41.21	0.000
GO	1	0.000602	8.39%	0.000602	0.000602	6.77	0.025
Error	11	0.000978	13.64%	0.000978	0.000089	0.48	0.821
Total	14	0.007172	100.00%				
Delamination factor (OUT)							
Model	3	0.010050	91.38%	0.010050	0.003350	38.85	0.000
Linear	3	0.010050	91.38%	0.010050	0.003350	38.85	0.000
S	1	0.002271	20.65%	0.002271	0.002271	26.34	0.000
F	1	0.006498	59.08%	0.006498	0.006498	75.35	0.000
GO	1	0.001280	11.64%	0.001280	0.001280	14.85	0.003
Error	11	0.000949	8.62%	0.000949	0.000086	0.79	0.673
Total	14	0.010998	100.00%				
Surface roughness (R_a)							
Model	3	0.269325	96.96%	0.269325	0.089775	116.89	0.000
Linear	3	0.269325	96.96%	0.269325	0.089775	116.89	0.000
S	1	0.035113	12.64%	0.035112	0.035112	45.72	0.000
F	1	0.183012	65.89%	0.183013	0.183013	238.29	0.000
GO	1	0.051200	18.43%	0.051200	0.051200	66.66	0.000
Error	11	0.008448	3.04%	0.008448	0.000768		0.102
Total	14	0.277773	100.00%				

Table 5.2. (Continued)

Source	DF	Seq SS	Contribution	Adj SS	Adj MS	F-Value	P-Value
Surface roughness (R_z)							
Model	3	10.5873	93.93%	10.5873	3.52911	56.76	0.000
Linear	3	10.5873	93.93%	10.5873	3.52911	56.76	0.000
S	1	4.5602	40.46%	4.5602	4.56020	73.34	0.000
F	1	5.6616	50.23%	5.6616	5.66161	91.05	0.000
GO	1	0.3655	3.24%	0.3655	0.36551	5.88	0.034
Error	11	0.6840	6.07%	0.6840	0.06218		0.456
Total	14	11.2713	100.00%				
Circularity error (C_e)							
Model	3	0.154127	98.49%	0.154127	0.051376	239.57	0.000
Linear	3	0.154127	98.49%	0.154127	0.051376	239.57	0.000
S	1	0.105111	67.17%	0.105111	0.105111	490.14	0.000
F	1	0.047740	30.51%	0.047741	0.047741	222.62	0.000
GO	1	0.001275	0.81%	0.001275	0.001275	5.95	0.033
Error	11	0.002359	1.51%	0.002359	0.000214		0.301
Total	14	0.156486	100.00%				

Quadratic mathematical models are employed to develop a correlation between the independent and dependent process parameters. Equations (5.1) through (5.7) display the results of the RSM-generated regression analysis:

$$T_h = 25.62 - 0.00183 \times S + 0.3649 \times F + 13.04 \times GO + 0.000004 \times S^2 - 0.000354 \times S^2 - 0.712 \times GO^2 - 0.000110 \times S^2 - 0.00525 \times S \times GO + 0.0254 \times F \times GO \quad (5.1)$$

$$T = 0.0850 - 0.000048 \times S + 0.000406 \times F + 0.0037 \times GO + 0.00250 \times GO^2 + -0.000031 \times F \times GO \quad (5.2)$$

$$Fd_{IN} = 0.9313 + 0.000028 \times S + 0.000497 \times F + 0.0407 \times GO - 0.00709 \times GO^2 - 0.000003 \times S \times GO + 0.000009 \times F \times GO \quad (5.3)$$

$$Fd_{OUT} = 0.9643 + 0.000035 \times S + 0.000563 \times F + 0.0086 \times GO - 0.000001 \times F^2 - 0.00097 \times GO^2 - 0.000004 \times S \times GO + 0.000088 \times F \times GO \quad (5.4)$$

$$R_a = 1.2897 + 0.000083 \times S + 0.001891 \times F + 0.08000 \times GO \quad (5.5)$$

$$R_z = 3.829 - 0.000944 \times S + 0.01052 \times F + 0.2137 \times GO \quad (5.6)$$

$$C_e = 0.3614 - 0.000143 \times S + 0.000966 \times F + 0.01263 \times GO \quad (5.7)$$

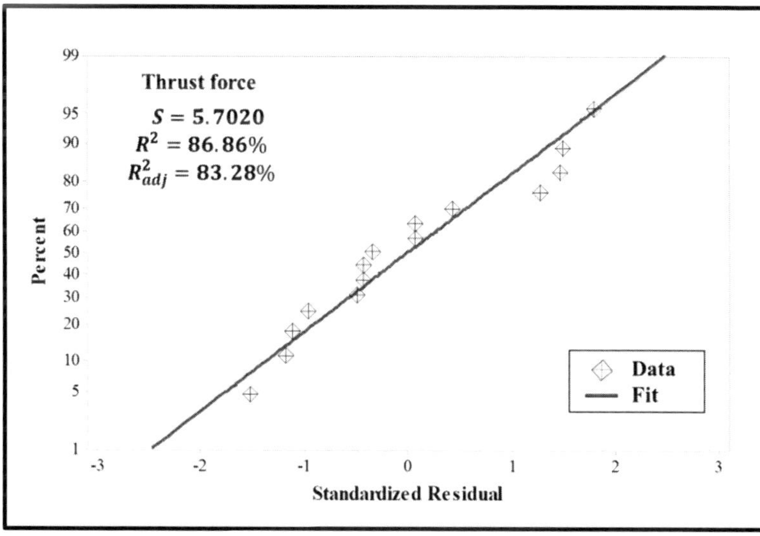

Figure 5.1. Probability plot for thrust force (T_h).

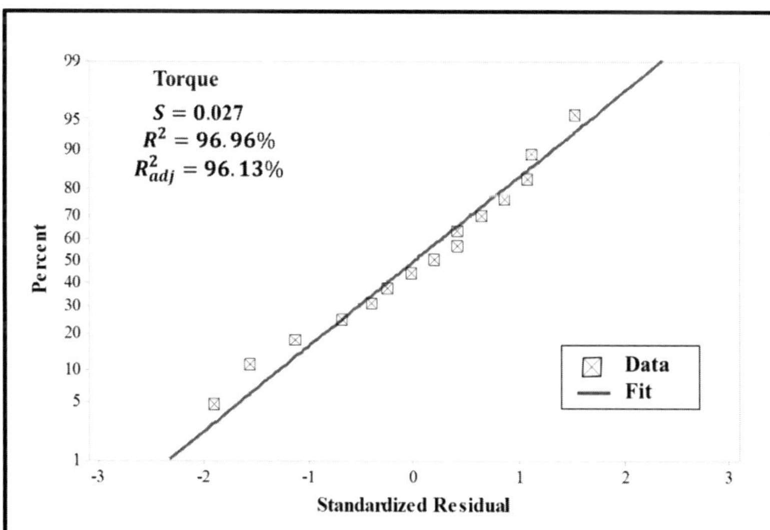

Figure 5.2. Probability plot for torque (T).

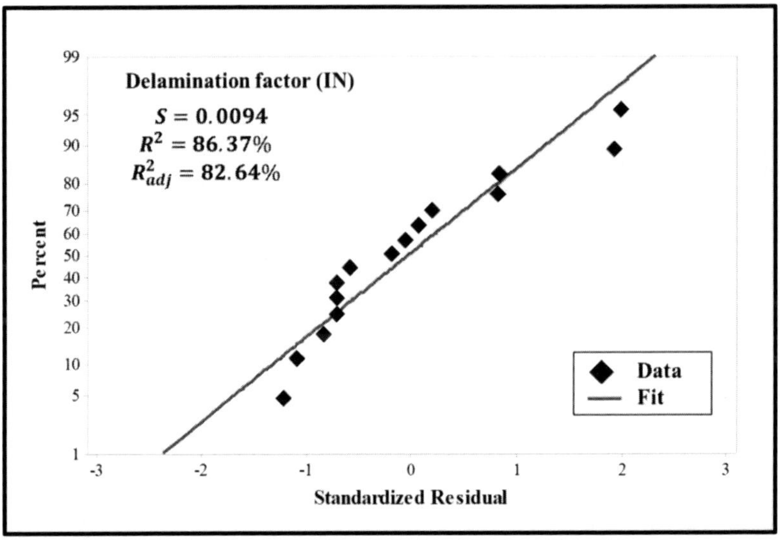

Figure 5.3. Probability plot for delamination factor (IN).

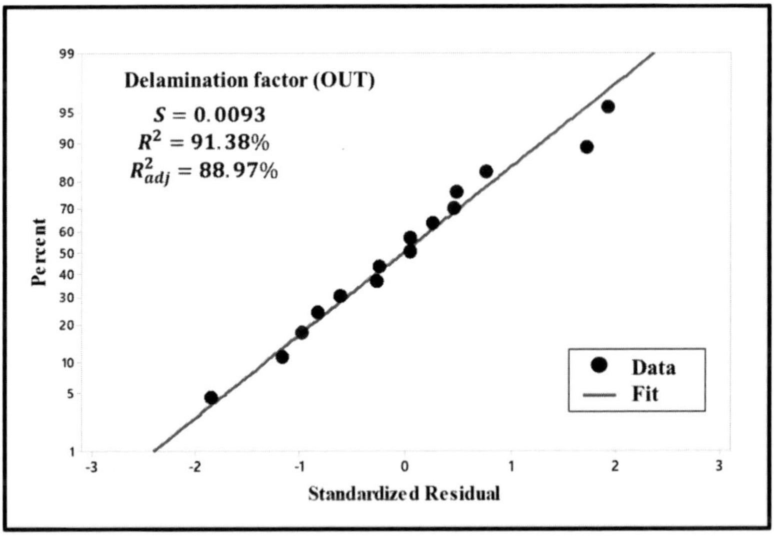

Figure 5.4. Probability plot for delamination factor (OUT).

Figures 5.1–5.7 show the drilling reaction. These are standardized probability plots that include Th, T, Ra, Rz, circulatory error (Ce) residual plots, and delamination factor (In & Out). A little outlier is visible on the near-to-fit line, which shows that the data is relatively normal. There is no discernible pattern or irregularity in its structure. The highly desired experimental answers are used to accomplish the recommended model predictions.

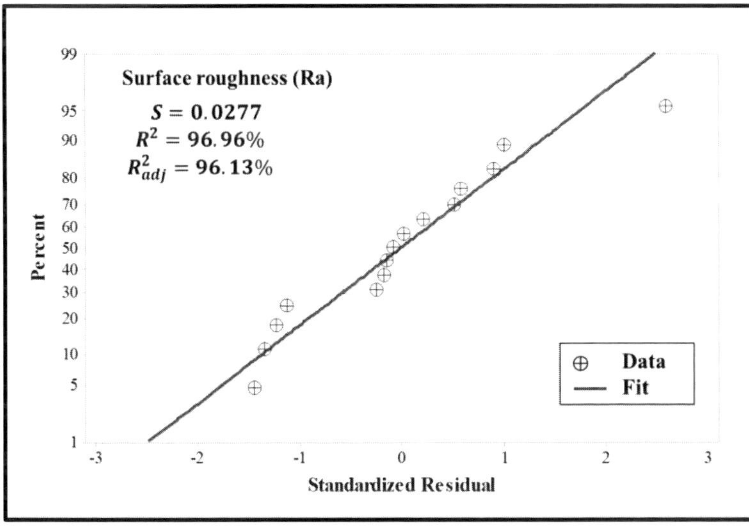

Figure 5.5. Probability plot for surface roughness (R_a) (J. Kumar & Kumar Verma, 2021).

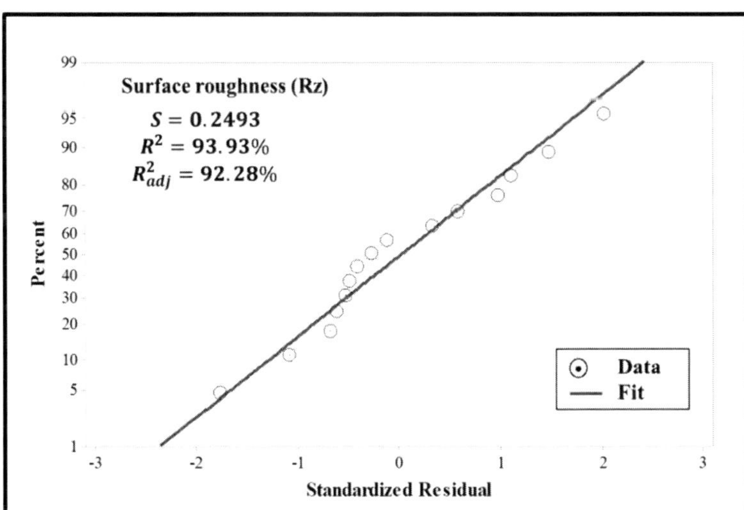

Figure 5.6. Probability plot for surface roughness (R_z) (J. Kumar & Kumar Verma, 2021).

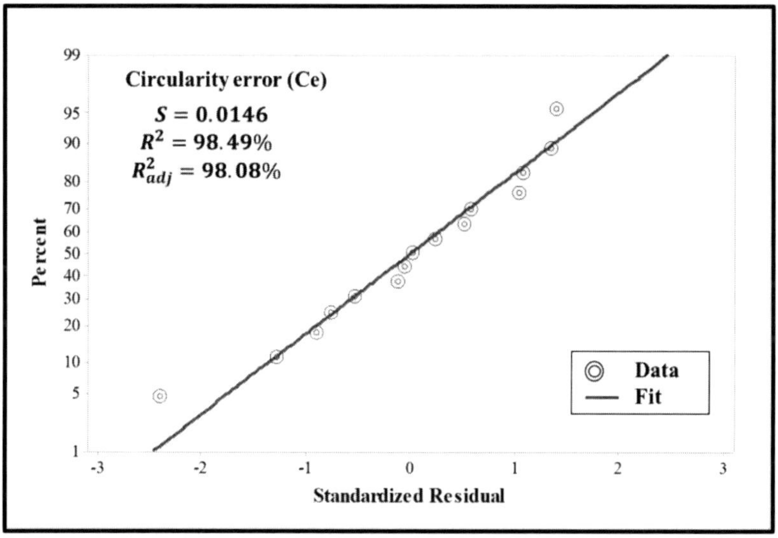

Figure 5.7. Probability plot for circularity error (C_e) (J. Kumar & Kumar Verma, 2021).

5.2.2. Mathematical Modeling of Milling Parameters

In this work, the second primary machining process is considered as the milling operation of proposed samples. The varying constraints were considered during experimentation, and accordingly, a Taguchi-based experimental array was designed. ANOVA was implemented to compute mathematical modeling and evaluate significant parameters. All the milling tests considered a 95% confidence level to estimate the output function value. Statistical analysis is a strategy used to analyze data and determine the relationship between parameters and their impact on responses. The assessment module's actual performance can be identified using this statistical method. It performs a comparison of the mean square value, F value, and P-value to assess experimental errors and allows for a 95% confidence level (Premnath et al., 2014). The section examines the significant effect of cutting speed, feed rate, and wt.% of GO on the milling response, namely MRR, cutting force, surface roughness, and delamination. Table 5.3 indicates that the contribution of cutting speed is the most significant factor affecting the value of 70.50% for MRR, 41.51% for cutting force, 84.53% for Ra, and 7.28% for the delamination factor. The feed rate is another influencing factor with the value of 15.26% for MRR and 8.03% for cutting force, 5.27% for R_a, and 62.60% for delamination. The graphene oxide wt.% is also a significant influencing factor contributing to 5.53% for cutting force and 9.79% for the delamination factor (Bhopale et al., 2017). This model shows that the contribution parameter error is significantly less (9.11%, 11.31%, 3.89%, and 18.67%). The adequacy of the model is R^2=90.89%, 88.69%, 96.11%, and 81.33% for MRR, F_c, R_a, and F_d, showing that unity is getting closer to an agreement (Bhattacharya et al., 2009; Patel et al., 2009).

Table 5.3. ANOVA for CNC milling

Source	DF	Seq SS	Contribution	Adj SS	Adj MS	F-Value	P-Value
Material removal rate (MRR)							
Regression	4	877.825	90.89%	877.825	219.456	12.47	0.008
S	1	680.970	70.50%	1.301	1.301	0.07	0.797
F	1	147.437	15.26%	42.595	42.595	2.42	0.181
D	1	42.167	4.37%	28.206	28.206	1.60	0.261
GO	1	7.252	0.75%	7.252	7.252	0.41	0.549
Error	5	88.026	9.11%	88.026	17.605		
Total	9	965.851	100.00%				
Cutting force (F_c)							
Regression	4	8242.8	88.69%	8242.83	2060.71	9.81	0.014
S	1	3858.2	41.51%	596.24	596.24	2.84	0.153
F	1	746.4	8.03%	74.67	74.67	0.36	0.577
D	1	3124.2	33.62%	2104.45	2104.45	10.01	0.025
GO	1	514.1	5.53%	514.08	514.08	2.45	0.179
Error	5	1050.8	11.31%	1050.81	210.16		
Total	9	9293.6	100.00%				
Surface roughness (Ra)							
Regression	4	16.4263	96.11%	16.4263	4.10659	30.90	0.001
S	1	14.4476	84.53%	0.2833	0.28327	2.13	0.204
F	1	0.9013	5.27%	0.1357	0.13567	1.02	0.359
D	1	1.0723	6.27%	1.0104	1.01044	7.60	0.040
GO	1	0.0052	0.03%	0.0052	0.00516	0.04	0.851
Error	5	0.6645	3.89%	0.6645	0.13291		
Total	9	17.0909	100.00%				
Delamination factor (F_d)							
Regression	4	0.360317	81.33%			4.36	0.092
S	1	0.032267	7.28%			1.56	0.280
F	1	0.277350	62.60%			13.41	0.022
D	1	0.007350	1.66%			0.36	0.583
GO	1	0.043350	9.79%			2.10	0.221
Error	4	0.082706	18.67%				
Total	8	0.443022	100.00%				

To establish a correlation between the independent and dependent process parameters, quadratic mathematical models are employed. Eqs. (5.8-5.11) shows the non-linear regression analysis:

$$MRR = -26.95 + 0.02400 \times C_v + 0.01456 \times F + 7.055 \times D + 16.17 \times GO - 0.000008 \times S^2 + 0.000026 \times F^2 - 2.130 \times D^2 - 3.927 \times GO^2 \quad (5.8)$$

$$F_c = 41.14 - 0.09894 \times C_v - 0.2003 \times F - 8.535 \times D + 61.05 \times GO + 0.000028 \times S^2 + 0.000517 \times F^2 + 20.87 \times D^2 - 13.18 \times GO^2 \quad (5.9)$$

$$R_a = 0.8822 - 0.001010 \times C_v - 0.000576 \times F + 3.321 \times D - 0.5628 \times GO + 0.000003 \times F^2 - 1.411 \times D^2 + 0.1072 \times GO^2 \quad (5.10)$$

$$F_d = 0.982 - 0.00584 \times C_v + 0.002687 \times F + 0.070 \times D + 0.0850 \times GO \quad (5.11)$$

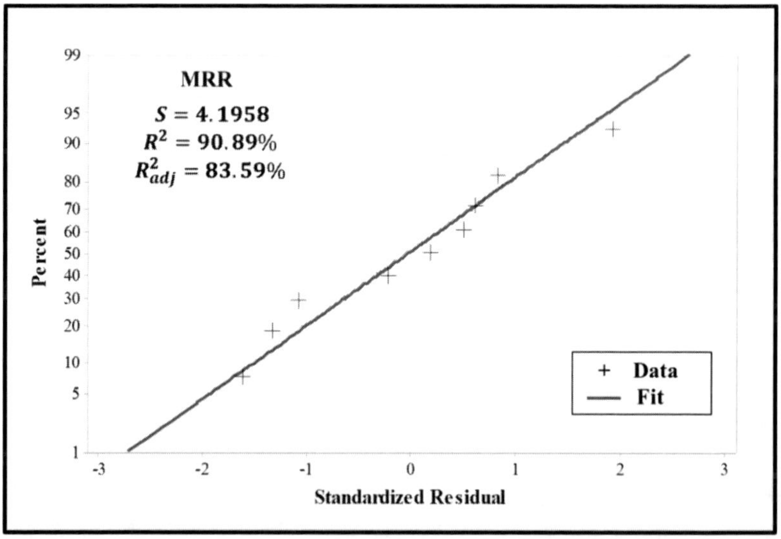

Figure 5.8. Probability plot for material removal rate (MRR).

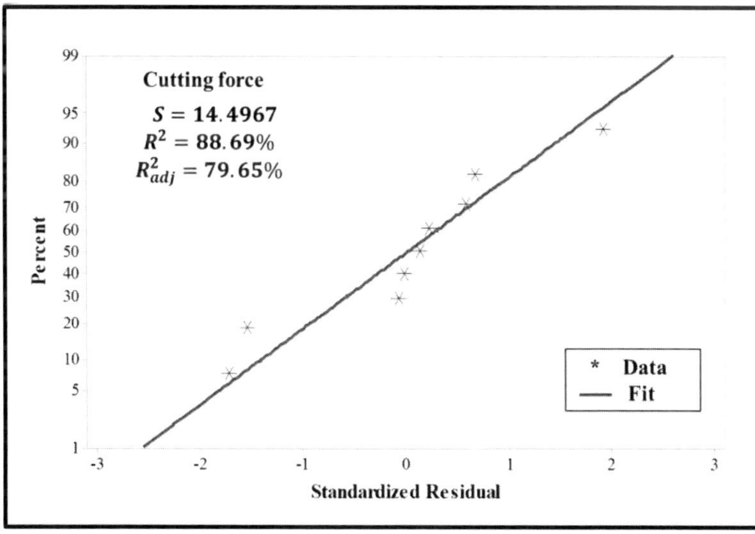

Figure 5.9. Probability plot for cutting force (F_c).

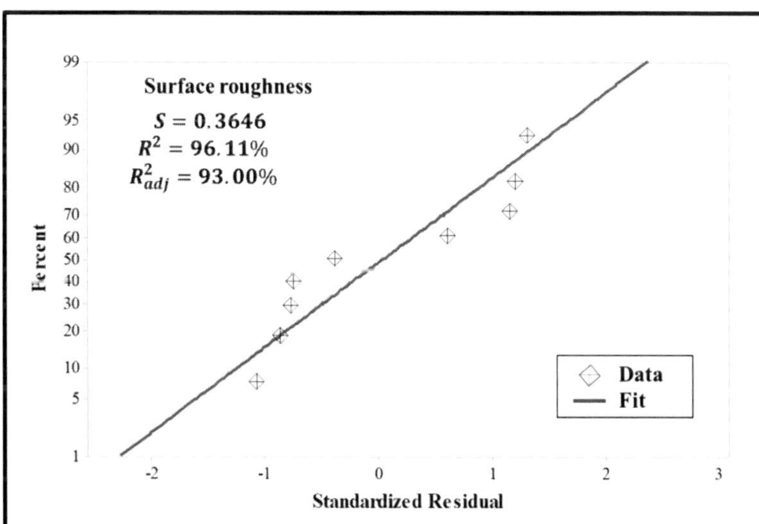

Figure 5.10. Probability plot for milling slot surface roughness (R_a).

All of the residual plot's points are located close to the mean line, or straight line, as seen in Figures 5.8 to 5.11, indicating that the data is both normally distributed and slightly deviates from the norm. It further noted that there is no discernible pattern or distinctive organization in the data. All of the data points on the standard plot are quite close to the mean line, which means the data is normally distributed but slightly off the norm. Further evidence that the mistakes are typically distributed is the fact that the residuals are smaller when plotted on a straight line. This residual data does not exhibit any anomaly in the ordinary probability plots, but there is a slight non-normality bias in the probability plot (Sahoo et al., 2017; Sarikaya & Güllü, 2014).

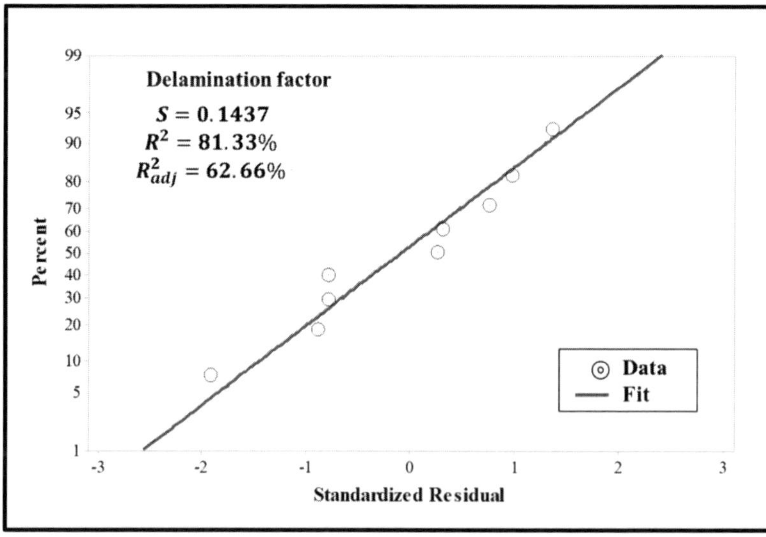

Figure 5.11. Probability plot for milling induced delamination factor (F_d).

5.3. Experimental Design

Over the past 20 years, prominent researchers have made significant advancements in the field of polymer machining. However, existing data show that very little work is available on the modeling and optimization of nanocomposite machining aspects. Furthermore, the lack of research on multi-objective optimization for machining performance has been observed. A thorough investigation is conducted into how different factors and cutting circumstances affect the performance of the machining process. Developments in the machining process can be completed through appropriate modeling and optimization approaches. Therefore, in order to regulate the process parameters, it is necessary to implement an appropriate optimization tool for machining conditions. Kevlar/glass/carbon/aramid-reinforced polymer composites are among the macro composites to which numerous distinguished researchers have made significant contributions to the improved effectiveness of machine outcomes. AHP, TOPSIS, Grey, utility functions, desirability functions, and other optimization modules aggregate process responses into a single objective function during multi-response optimization. AHP provides preferable consistency and addresses the criteria for subjective judgment decisions. It is unrealistic for most investigations to assume that when conflicting results are aggregated, there will be no response association and the same priority weight. The results are ambiguous, inaccurate, and the result of these theories. This chapter's examination of the PCA module skillfully challenges these fundamental and conventional beliefs. By turning the correlated output functions into separate values, PCA solves the multicriteria optimization problem.

The process of response's established statistical weight is estimated using PCA. In conventional manufacturing processes, the optimization method is widely applied to evaluate the best combinations of machine constraints for contradicting machining response studies. In addition, renowned academics utilize it extensively for cutting-edge manufacturing processes like welding (S. Kumar & Singh, 2019), casting (B. Das et al., 2016), milling (Uzun, 2019), turning (Gok, 2015), etc. The PCA approach is a helpful statistical strategy that is mainly applied to response priority weights with contradictory characteristics. Through parametric appraisal, it is possible to make the suggested material better at being machined while also cutting down on damage and waste (Sorrentino et al., 2017). Additionally, there is a lack of information regarding the machining behavior of nanocomposites. It needs more attention from academia and industry. Numerous distinguished academics thoroughly analyzed the delamination and other machining-induced damage ideas (An et al., 2014; Qiu et al., 2018; Xu et al., 2013, 2018, 2019b, 2020). Manufacturing sectors, on the other hand, continue to be too concerned about delamination, which directly impacts productivity and quality issues. Therefore, it is essential to investigate this complicated matter in GO/CF reinforced polymer nanocomposites employing efficient instruments and optimization methodologies. The technical, financial, and environmental needs are all considered while optimizing multicriteria optimization modules. Improving the quality and productivity indices of machining processes is greatly aided by the optimization of various machine limitations (Ali et al., 2019). Fiber-reinforced polymers (FRP) are typically used in the aerospace, automotive, marine, and construction component manufacturing industries to fulfill the requirements of precision component assembly, near tolerances, weight reduction, and production simplicity. Important problems like fiber pull out, delamination, plastic deformation, matrix debonding, etc., arise when producers machine epoxy composites (Panchagnula & Palaniyandi, 2018). It produces a challenging environment for epoxy composites' final assembly and manufacture. Due to their anisotropic performance and non-homogeneity, epoxy composites require significantly different machining techniques than metal and its alloys. Understanding the machinability features and optimizing machining performance is essential for the right use of any innovative composite. As a result, industry, research institutions, and academia have latent research interests in machining these new composites.

Lean, agile, and flexible manufacturing environments are used by various polymer businesses in today's rapidly expanding manufacturing sectors to adapt to their clientele's constantly shifting demands and wants. Customer satisfaction is crucial for any organization or enterprise regarding procedures or goods. It is possible to meet the drastically shifting client demands by using an automated, computer-controlled manufacturing system. There is a high machining cost, expert operator cost, and setup/installation cost associated with these precise machines. From these configurations, the manufacturer is confronted with an extremely difficult payback cost. The limitations and drawbacks of existing methods include negligible response correlation and weight assignment responses primarily dependent on decision-makers (DM). Fewer experiments were conducted on drilling polymer nanocomposites, according to a thorough literature review of the preceding study. Eminent academicians have been observed to address the manufacturing and machinability aspects of polymer composites extensively (Baraheni & Amini, 2019; D. Kumar & Singh, 2019; Palanikumar, 2011). The researchers investigate tool wear, fiber pullout, matrix debonding, and other factors, and as a result, they have developed a variety of robust optimization tools (Coleman et al., 2006; Rajkumar et al., 2017). Most studies have focused on individual optimization modules; consequently, there is a scarcity of research on the hybridization strategies of various optimization modules during the machining of epoxy nanocomposites. It is believed that the machining characteristics can be significantly improved by implementing appropriate parametric parameters using hybridization modules.

5.3.1. Response Surface Methodology (RSM)

The Box Behnken design (BBD) simultaneously considers all variables/factors at maximum and minimum levels. Each variable was concurrently selected at its maximum and lowest level in the BBD. Several distinct inputs and outputs (responses) were correlated, which was determined by creating a polynomial equation using RSM. So, compared to the experiment's conventional design, it greatly minimizes the number of tests runs and improves cost-effectiveness. A second-order polynomial regression for the quadratic model is represented by the following equation.

$$F\ (X) = \beta_0 + \sum_{j=1}^{n} \beta_j X_j + \sum_{j-1}^{n} \beta_{jj}\, X_j^2 + \sum_{i=1}^{n=1} \sum_{j=2}^{n} \beta_{ij} X_i X_j \qquad (5.12)$$

Here β_j linear, β_{jj} quadratic, and β_{ij} interaction terms.

5.3.2. Taguchi Theory

Traditional optimization techniques hold the other components fixed while considering the changing constraint of one of the factors. However, the experimental statistical design method offers a more convenient and effective alternative to optimize several operational variables. Numerous evolutionary optimization modules and Taguchi-based techniques are part of the competent test design approach. (Islam & Pramanik, 2016; Kulkarni et al., 2020; Sampath Kumar et al., 2017). A Taguchi array is a statistically balanced experimental array that effectively shortens test runs. It is a highly regarded experimental pattern that is chosen for case studies with variable factors, decision science, operational research, multicriteria evaluation, and machining test design, among other applications. Asghar et al., (Asghar et al., 2014) investigated a comparative study of Taguchi array and central composite design (CCD) to estimate the feasibility and efficacy of the experimental array. The outcomes of the interaction plot and predicted value show the feasibility of the Taguchi array and are also cost-effective. Myers et al., (Raymond H. Myers, Douglas C. Montgomery, 2016) examined a comparison between Taguchi and CCD for a case study. They discovered that, in contrast to CCD, Taguchi created a favorable influence and improved quality with statistics. Thus, the present work explores the Taguchi method-optimized GO/CF polymer nanocomposite milling technique. The Taguchi idea was used to plan the experiment and find the relationship between the input parameters that change a process parameter and the output value (Palanikumar et al., 2006; Shunmugesh & Panneerselvam, 2016, 2017). Finally, the optimal condition was obtained by the S/N ratio. For MRR, the maximization function of higher the better (HB) was treated by using the following equation:

$$\left({}^{S}/_{N}\right)_{ij} = -10\log\left(\frac{1}{n}\sum_{j=1}^{n}\frac{1}{x_{ij}^{2}}\right) \tag{5.13}$$

The Fc and Ra desired consider the lower the better (LB) criteria by the below equation:

$$\left({}^{S}/_{N}\right)_{ij} = -10\log\left(\frac{1}{n}\sum_{j=1}^{n}x_{ij}^{2}\right) \tag{5.14}$$

5.3.3. Principal Component Analysis (PCA)

The principal component analysis (PCA) approach was developed by Karl Pearson in 1901. It comprises a fundamental statistical method to assess the major principal component (PC) (Karl Pearson F. R. S., 1901). Principal components (PCs), which are linearly uncorrelated factors, have been altered using orthogonal transformation. PCA, in conjunction with the CoCoSo module, offers more control over quality assurance and productivity issues related to manufacturing processes. (Abhishek et al., 2014). In this analysis, the correlation factor and the specific response weight are evaluated.

A. Calculate the Correlation Coefficient

$$R_{ij} = \frac{Cov\,\{X_i(k), X_i(l)\}}{\sigma_{(xi)}(k) \times \sigma_{(xi)}(l)} \tag{5.15}$$

where I = 1,2,3,4.................n and j = 1,2,3,4................ n, $X_i(j)$, is normalized for each response, $Cov\,\{X_i(k), X_i(l)\}$ is the covariance of response variables, $\sigma_{(xi)}(k) \times \sigma_{(xi)}(l)$ are the standard deviation of the response variable.

B. Eigenvalue and eigen vector

$$(R_{ij} - \gamma_k I_m)V_{ik} \tag{5.16}$$

where, γ_k= eigen values and k= 1,2,3,4……. n ,

Thus, the PC values are

$$Y_{mk} = \sum_{i=1}^{n} x_m(i)V_{ik} \tag{5.17}$$

where, $Y_{m1}, Y_{m2}, \ldots$ PC value…etc. The first major element, $Y\text{-}m1$, has the highest variance of the data since the main components are adjusted for variance in decreasing order.

While correlated response is often overlooked in standard optimization modules, it is covered in the optimization module employed in this study. Principal components (PCs) are uncorrelated measures of machining performance that result from correlated responses becoming independent. When using multicriteria decision-making (MCDM), the CoCoSo module is regarded as a reliable statistical instrument for resolving important problems. According to state-of-the-art specifications, many modules and formulas for aggregating multiple features exist. This study explores a noteworthy concept regarding combining several aspects in machining. Complex optimization problems can be effectively solved by using the CoCoSo-PCA coupling concept, which offers practical feasibility.

5.4. Multi-Objective Optimization

A new optimization technique called CoCoSo is a reliable statistical instrument for determining a varied response. This approach was first suggested by Yazdani et al., (Yazdani et al., 2019) in order to assess specific supply chain management criteria in France (2018). Nowadays, it is widely used for optimization in the banking industry, manufacturing sectors, commercial sector, and power control (Karaşan & Bolturk, 2019; Mi & Liao, 2020; Wen et al., 2019) for DM. This study optimizes the parametric optimization and drilling and milling machinability of the GO/CF/polymer composites using a combined CoCoSo approach. The CoCoSo approach normalizes the values in accordance with the intended quality standards as advantageous or economical. Based on the WASPAS methodology, the weighted comparability sequence and the weighted quantity of machining response for each alternative were determined. Subsequently, each alternative is referenced by three aggregation operators (Kia, Kib, and Kic), which generates three compromise scores. The suggested methodology for compromise solutions consists of the statistically proven result and the exponentially weighted product model. After assessing the alternatives and the corresponding criteria, CoCoSo employs the subsequent procedures to resolve the multifunctional decision problem (Kharwar et al., 2020).

Step-1: DM is stated as:

$$x_{ij} = \begin{bmatrix} x_{11} & x_{12} & \cdots & x_{1n} \\ x_{21} & x_{22} & \cdots & x_{2n} \\ \cdots & \cdots & \cdots & \cdots \\ x_{1m} & x_{2m} & \cdots & x_{mn} \end{bmatrix} \tag{5.18}$$

Here, x_{ij}=MD and m = 1,2,3 & n = 1,2,3.

Step-2: The intended quality control criteria are used to normalize the attribute response. To reduce the uncertainty, the CoCoSo module initially normalizes the measured values, a procedure known as data pre-processing. According to this data, the original sequence has been moved to the same range between zero and one. The primary factor influencing data pre-processing methods is the desired standard of function, which is referred to as "cost criteria."

$$r_{ij} = \frac{\min x_{ij} - x_{ij}}{\max x_{ij} - \min x_{ij}} \tag{5.19}$$

where x_{ij}= selected data, $\min x_{ij}$ Minimum, $\max x_{ij}$ Max data in array

Step 3: Sum of product with proportionate component weighted at a normalized value for each alternative is total weight comparability sequence (Si). At each alternative normalization value, the comprehensive comparability weight (Pi) assessment summarizes the exponential proportional component weighted. The product sum's weighted multiplier and the WASPAS method's exponential sum can be employed to determine the Si and Pi values.

$$S_i = \sum_{j=1}^{n} (\mathrm{w}_1 \times N\mathrm{R}_a + \mathrm{w}_2 \times N\mathrm{R}_z + \mathrm{w}_3 \times NC_e) \tag{5.20}$$

$$P_i = \sum_{j=1}^{n} ((N\mathrm{R}_a)^{\mathrm{w}_1} + (N\mathrm{R}_z)^{\mathrm{w}_2} + (NC_e)^{\mathrm{w}_3}) \tag{5.21}$$

Step-5: By using three evaluation criteria and the following averaging approaches, the relative weights of the solutions are obtained.

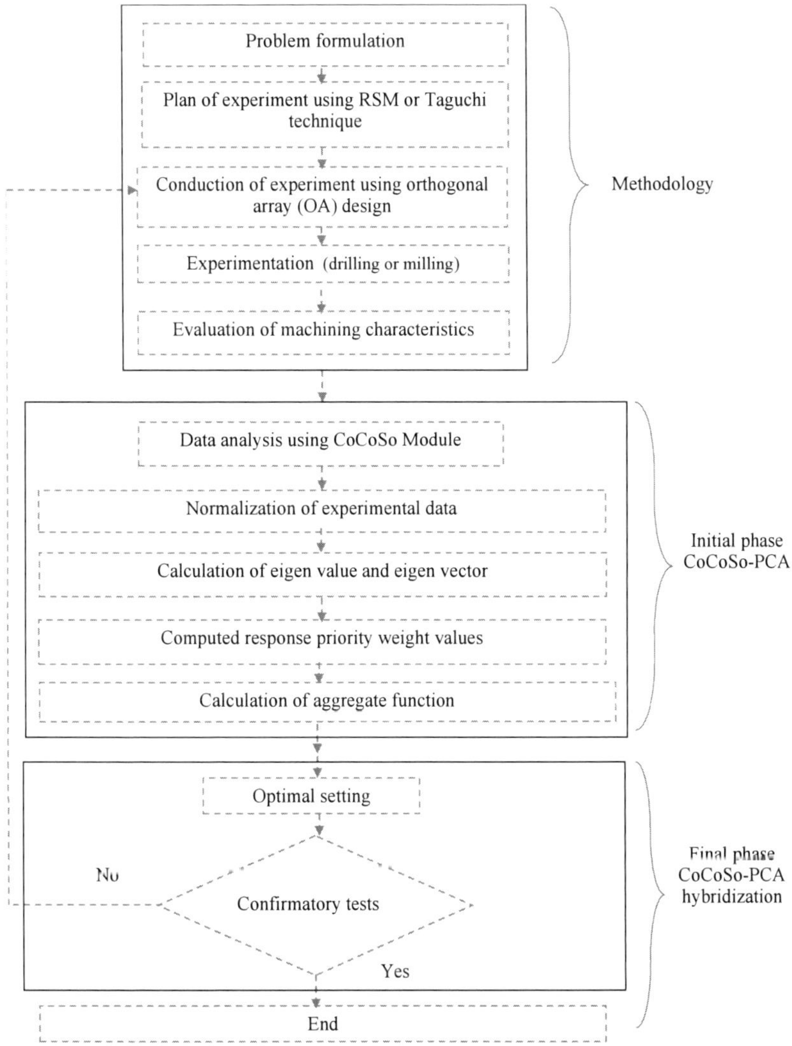

Figure 5.12. Method for CoCoSo-PCA (J. Kumar & Kumar Verma, 2021).

$$K_{ia} = \frac{P_i + S_i}{\sum_{i=1}^{m}(P_i + S_i)} \tag{5.22}$$

$$K_{ib} = \frac{S_i}{min\, S_i} + \frac{P_i}{min\, P_i} \tag{5.23}$$

$$K_{ic} = \frac{\lambda(S_i) + (1-\lambda)(P_i)}{(\lambda \max S_i + (1-\lambda)\max P_i)} \tag{3.24}$$

Here, $\lambda = 0.5$ (Yazdani et al., 2019).

Step 6: The final ranking of the alternatives is determined by the Ki values.

$$K_i = (k_{ia}k_{ib}k_{ic})^{\frac{1}{3}} + \frac{1}{3}(k_{ia} + k_{ib} + k_{ic}) \tag{5.25}$$

This methodology processed the compromise ranking results, integrating the PCA module to include the compromise values that suited each alternative to resolve multicriteria problems. Figure 5.12 illustrates the various steps taken to optimize the machining process parameters.

5.4.1. Case Study 1: Parametric Optimization Approach Drilling Process

The CoCoSo-PCA method was utilized in this study to evaluate and optimize drilling data, and RSM-designed-based experiments were conducted. Several drilling process limitations show the direct and indirect causes of the machining. To gain a more comprehensive understanding of the machining environment, it is possible to select a variety of process parameters, including the diameter of the cutter, the type of material coated on the tool, the tip angle, and the type of coolant (Fu et al., 2018; Sridharan et al., 2016; Thirukkumaran et al., 2020). Additionally, the steel industry had previously taken certain process characteristics into account (Senthilkumar et al., 2014), laser drilling of Inconel-718 (Tam et al., 1993). It has been noted in the literature that machining qualities can be influenced by a number of different parameters. However, there are constraints on how many parameters can be considered and how many responses can be processed simultaneously in experiments; doing differently might impact machining time, experimentation cost, and complexity. This study follows the well-established RSM concept of 15 experiments, which identifies three process constraints: spindle speed (S), feed rate (F), and weight percent graphene oxide (GO). Previous state-of-the-art pilot tests were used to select limitations. Three distinct levels of variation are applied to the three restrictions. A total of fifteen distinct experiment sets were carried out. After that, suitable methodology and equipment were utilized to evaluate the machining performance. According to Table 5.4, the drilling was done using the RSM-based Box Behnken design (BBD). Responses for the optimization procedure were Ra, Ce, and surface roughness on a 10-point scale (Rz).

The drilling of GO/CF-reinforced composites is optimized regarding process parameters and machinability. PCA assigns the priority weight of the response, and the combined compromise solution aggregates the multiple responses. The uniform response weight is assumed by most pioneer academicians during the aggregation of responses. A hypothesis like this leads to uncertainty and inaccuracy in the results, ultimately affecting the module's efficiency. The proposed hybrid module optimized the conflicting machining response during drilling experiments. All values were initially normalized within the range of 0 to 1. Using Equation (5.19) and the related values listed in Table 5.4, the normalization approach is considered a cost criterion for the machining replies. Eq. (5.16) was applied to the normalized decision matrix to get the data eigenvector and eigenvalue. Eq. (5.17) has been employed to evaluate the PCA weight, and Figure 5.13 illustrates the eigenvalues and corresponding contributions. Calculate the Si and Pi values by dividing the observed values by the projected value after the weight calculation of the normalized value for each experimental test. Table 5.4 presents the three constant evaluations Kia, Kib, and Kic of each experiment, which are tested using Eqs. (5.22-5.24) to estimate the relative weight criteria. The comparability weight (Pi) data and the selected comparability sequence (Si) were employed to generate the relative weight attribute matrix.

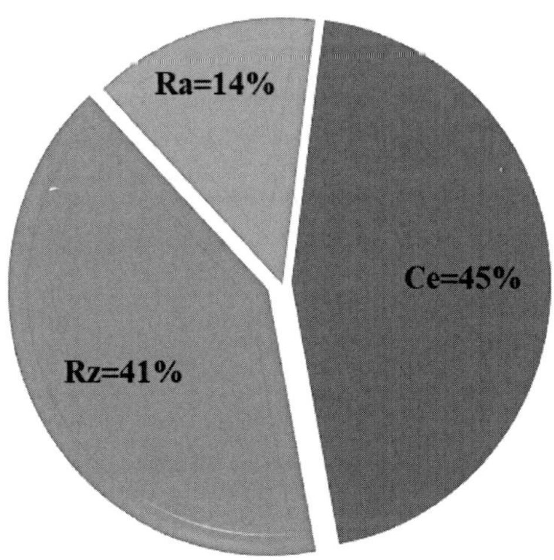

Figure 5.13. Contribution for PCs value (J. Kumar & Kumar Verma, 2021).

Table 5.4. Box Behnken (BB) and analyzed data (J. Kumar & Kumar Verma, 2021)

Exp. No.	S	F	GO	R_a	R_z	C_e	R_a	R_z	C_e	S_i	P_i	K_{ia}	K_{ib}	K_{ic}	K_i	Rank
	(rpm)	(mm/min)	(wt.%)	(µm)	(µm)	(mm)	(µm)	(µm)	(mm)							
1	800	80	2	1.69	4.2	0.366	0.911	0.567	0.398	0.545	2.447	0.073	13.559	0.767	15.314	7
2	2400	80	2	1.80	2.9	0.133	0.667	1	1	0.954	2.946	0.096	22.448	1	24.834	1
3	800	240	2	1.99	5.8	0.52	0.244	0.033	0	0.049	1.037	0.027	2	0.278	2.5507	15
4	2400	240	2	2.10	4.43	0.287	0	0.49	0.602	0.468	1.536	0.049	11.107	0.514	12.325	10
5	800	160	1	1.75	5.09	0.42	0.778	0.27	0.258	0.335	2.093	0.06	8.8978	0.623	10.271	12
6	2400	160	1	1.84	3.66	0.169	0.578	0.747	0.907	0.788	2.764	0.087	18.872	0.911	21.015	3
7	800	160	3	1.87	5.7	0.41	0.511	0.067	0.284	0.216	1.803	0.05	6.1791	0.518	7.2873	13
8	2400	160	3	2.09	3.76	0.21	0.022	0.713	0.801	0.654	2.365	0.074	15.725	0.774	17.54	5
9	1600	80	1	1.65	3.6	0.211	1	0.767	0.798	0.811	2.798	0.089	19.373	0.926	21.554	2
10	1600	240	1	1.95	5	0.37	0.333	0.3	0.388	0.34	2.118	0.06	9.0329	0.63	10.424	11
11	1600	80	3	1.78	3.7	0.25	0.711	0.733	0.698	0.715	2.686	0.083	17.296	0.872	19.332	4
12	1600	240	3	2.09	5.9	0.401	0.022	0	0.307	0.128	1.213	0.033	3.8046	0.344	4.5317	14
13	1600	160	2	1.88	4.4	0.32	0.489	0.5	0.517	0.505	2.4	0.071	12.697	0.745	14.39	9
14	1600	160	2	1.89	4	0.301	0.467	0.633	0.566	0.583	2.506	0.076	14.396	0.792	16.216	6
15	1600	160	2	1.90	4.3	0.31	0.444	0.533	0.543	0.525	2.426	0.072	13.124	0.757	14.849	8

The function value (Ki) was determined using Eq. (5.25). Table 5.4 describes the corresponding values. The results of Experiment No. 2 show that its Ki value is 24.834, which is very close to the choices in the orthogonal array.: spindle speed (S) of 2400 rpm, feed rate (F) of 80 mm/min, and graphene weight percentage (GO) of 2%. The hybrid CoCoSo-PCA approach's S/N ratio plot of the objective function value (Ki) shows the best constraint level in Figure 5.14.

With S of 2400 rpm, F of 80 mm/min, and GO weight percentage of 1% are the optimal conditions for the drilling test. It is evident that the objective function value is 25.696, which exceeds the array assessment value. Due to its intermittent cutting function, the drilling process is more intricate than other machining procedures. To induce material deformation, the temperature in the deformation region must be sufficiently elevated. To soften the cutting zone and make cutting easier, the cutting force should not exceed the cutting tool's capabilities. We have observed that the cutting temperature can increase as the cutting speed increases (Gong & Li, 2019). Due to the high spindle speed and cutting zone heat, material removal is smoothed., which reduces the surface roughness (Kalidass & Palanisamy, 2014). It may be regulated by keeping the feed rate at 80 mm/min and the spindle speed at 2400 rpm. The Ce is at its lowest at a reduced 80 mm/min feed rate. At the same time, there was a decrease in the weight percentage of GO in the fiber breakage and cracking because the polymer composite's laminar strength increased (Jenarthanan & Jeyapaul, 2018).

The graph showing the predictable results for CoCoSo-PCA is shown in Figure 5.14. Response diagrams were used to analyze how drilling parameters affected the results. Ra, Rz, and Ce are minimal with S at 2400 rpm, F at 80 mm/min, and graphene oxide wt. %. The parametric combination demonstrates that a mere 1% of GO can enhance the desirable drilling characteristics. For low values of circularity error and surface roughness, employing a combination of a low feed rate and a higher spindle speed is preferable. The roughness value increases as the feed rate increases. This is a consequence of the drill's increased force on the work sample, which can lead to a more substantial bending and beginning of a fracture (Köklü et al., 2019; Kyratsis et al., 2018; Wang et al., 2004). The matrix phase can be softened by high spindle speed, increasing the machining interface temperature (tool cutting edge and workpiece). Consequently, the cutting force must be reduced (Kaybal et al., 2019; Tsao, 2008). Less damage from drilling-related issues, such as matrix debonding and fiber breakage, is produced. In past studies conducted by eminent researchers, a comparable outcome and trend in machining performance were observed (Geier & Szalay, 2017; Ragunath et al., 2017). The suggested nanocomposite, which is very vulnerable to drilling-induced defects such as delamination, matrix debonding, and fiber breaking, comprises graphene oxide/carbon fiber laminates in an epoxy matrix. The addition of GO wt. %, and the selection of appropriate process parameters can overwhelm it.

The confirmatory test was conducted under the optimal conditions of a spindle speed (S) of 2400 rpm, a feed rate (F) of 80 mm/min, and a weighting percentage of GO of 1% of the objective function value. The results suggest a significant improvement in the response to machining. The values of Ra and Rz decreased from 1.80 to 1.55 μm and 2.90 to 2.50 μm, respectively. In addition, the Ce decreased from 0.133 to 0.100 mm. According to the optimal configuration, the feed rate value substantially impacts the improvement of surface finishing. This is due to a higher S and a reduced F can mitigate vibration and the formation of defects and cracks (Gopal et al., 2017; Tsao, 2009). The weight percentage in the machined sample condition also significantly improves the bonding strength. The objective function value of 33.946 confirms the feasibility of the proposed hybridization approach in a machining environment.

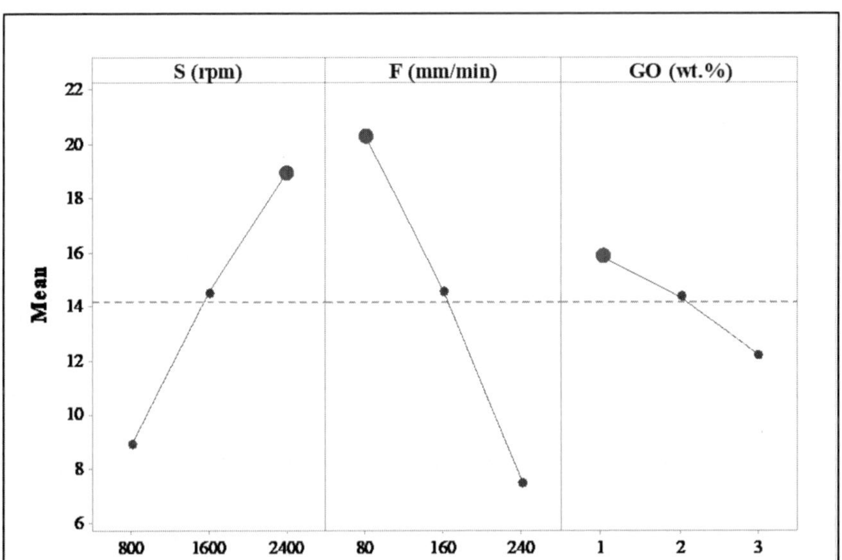

Figure 5.14. Optimal condition (J. Kumar & Kumar Verma, 2021).

5.4.2. Case Study 2: Parametric Optimization Approach Milling Process

In the case of a specific issue, the number of causes is a significant factor. The coordinated and uncomplicated approach employed in this method is frequently used to identify the underlying cause of the issue. In order to illustrate the prospective factors, both direct and indirect that contribute to the manufacturing process, Figure 5.15 illustrates the CED of the milling. (Hou et al., 2007; Rajesh Mathivanan et al., 2016; Yan & Li, 2013). Machining characteristics can be influenced by various other factors, as evidenced by Figure 5.15. It is impossible to examine all aspects and responses at once because it would increase machining time, experimentation expense, and complexity (Datta & Sankar Mahapatra, 2010; Selvarajan et al., 2015). The observed data has been analyzed and optimized using the Taguchi-based CoCoSo-PCA approach in this study. As a result, the L9 orthogonal array was implemented in this investigation to adjust the S, F, D, and wt.% of GO at three distinct levels as mentioned in Table 5.5.

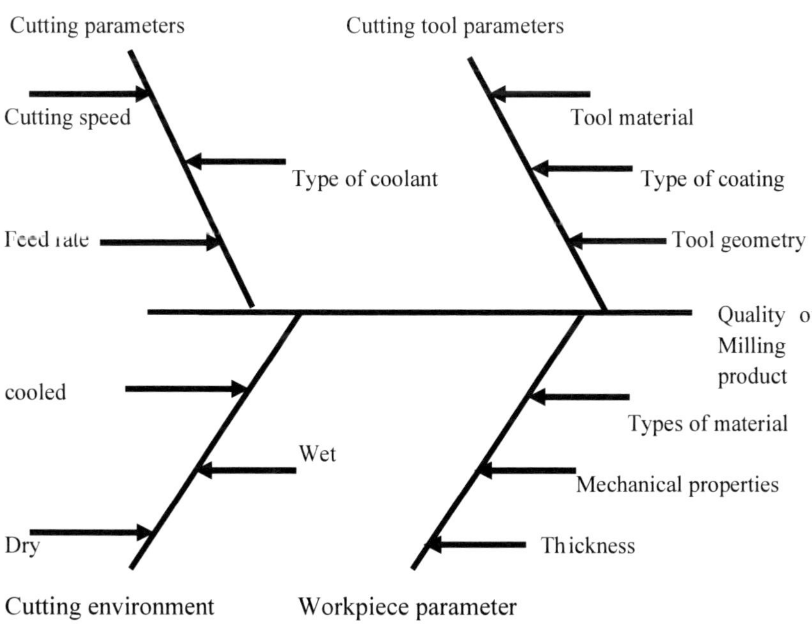

Figure 5.15. Cause and effect diagram (CED) (J. Kumar & Verma, 2021b).

This critical task of correlated responses can be investigated using the CoCoSo-PCA hybrid optimization strategy. Researchers have implemented this module in numerous optimization case studies, including power controllers, energy optimization, biological investigations, economics, commercial challenges, and other multicritical industrial engineering decision-making issues because of its reduced processing and complexity. Data pre-processing is the method by which the CoCoSo research first normalizes the observed data to lessen the variability across conflicting functions. Preprocessing data shows that the starting order should be moved to a comparable sequence to normalize the observed data from zero to one per the intended criteria. While MRR, Fc, and Ra are the characteristics of machining performance, S, F, D, and wt.% of GO are the process parameters taken into consideration in this study. The output is first turned into 0 and 1 using a normalization equation. Table 5.5 lists the normalized value for each response machining, which is determined using Eqs. (5.19). The PCA method follows, using Equations (5.16–5.17) to figure out how much each performance attribute is significant. The square of the eigenvalues represents a PCA's quality attributes proportion. Additionally, Table 5.5 shows that the percentage of the first PC value is greater at 42.18%.

The contribution quantities for MRR, Fc, and Ra are evaluated at 0.01, 0.54, and 0.45, respectively, as indicated in Figure 5.16. The percentage contributions for surface roughness, cutting force, and MRR are discovered to be 42.18%, 36.57%, and 21.25%, in that order. In conclusion, the total evaluation value is evaluated using Eq. (5.17). The eigen values are given out in decreasing order., as specified in the CoCoSo technique, and subsequently ranked. The highest Ki value observed was 8.4 for Exp. No. 6. The optimal machining parameters are determined by the L_9 OA for multicriteria features are S2-F3-D1-GO2, which is 25.12 m/min, 240 mm/min feed rate, 0.5 mm cutting depth, and 2 wt.% graphene oxide. According to HB criteria, the S/N ratio plot (Figure 5.17) indicates the optimal parametric settings as S2F2D1GO1 (S = 25.12 m/min, F = 160 mm/min, D = 0.5 mm, and GO% = 1%). The predicted value for Ki's S/N ratio using CoCoSo-PCA Taguchi's method is 9.06. The response table is derived from the Ki alternatives mean analysis is presented in Table 5.5. CoCoSo is significantly impacted by the D, with the highest delta value of 3.21. Following feed rate and GO, which have delta values of 0.76 and 0.29, respectively, cutting speed is the second variable that affects performance.

Table 5.5. Taguchi L_9 OA and analysis of data

Exp. No.	S	F	D	GO	MRR	F_c	R_a	Normalized value			*weight*		relative weights			*Ranking value*	Rank
	(m/min)	(mm/min)	(mm)	(wt.%)	(mm^3/min)	(N)	(μm)	MRR	F_c	R_a	S_i	P_i	K_{ia}	K_{ib}	K_{ic}	K_i	
1	12.56	80	0.5	1	3.79	15.78	1.12	0	0.8	0.7	0.75	1.74	0.1	4.41	0.62	5.77	5
2	12.56	160	1	2	11.78	42.57	1.48	0.66	0.32	0.39	0.35	2.19	0.1	2.89	0.64	4.2	7
3	12.56	240	1.5	3	11.18	60.07	1.39	0.61	0	0.47	0.22	1.7	0.07	2	0.48	2.97	9
4	25.12	80	1	3	10.78	17.75	1.31	0.57	0.76	0.54	0.66	2.62	0.13	4.53	0.82	6.26	3
5	25.12	160	1.5	1	12.37	16.78	1.48	0.7	0.78	0.39	0.6	2.53	0.12	4.22	0.79	5.87	4
6	25.12	240	0.5	2	15.97	5.60	0.78	1	0.98	1	0.99	2.99	0.16	6.25	1	8.4	1
7	37.68	80	1.5	2	9.18	53.67	1.54	0.44	0.12	0.34	0.22	1.92	0.08	2.13	0.54	3.21	8
8	37.68	160	0.5	3	4.59	9.53	1.02	0.07	0.91	0.79	0.85	2.82	0.14	5.5	0.92	7.45	2
9	37.68	240	1	1	7.58	4.73	1.93	0.31	1	0	0.54	1.98	0.1	3.61	0.63	4.95	6

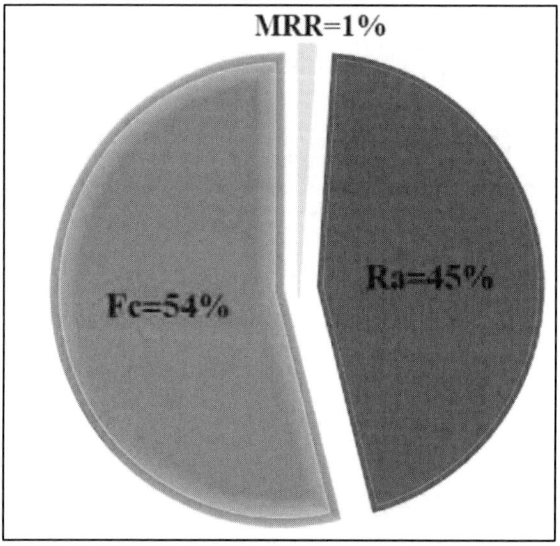

Figure 5.16. Contribution to the value of PCs (J. Kumar & Verma, 2021b).

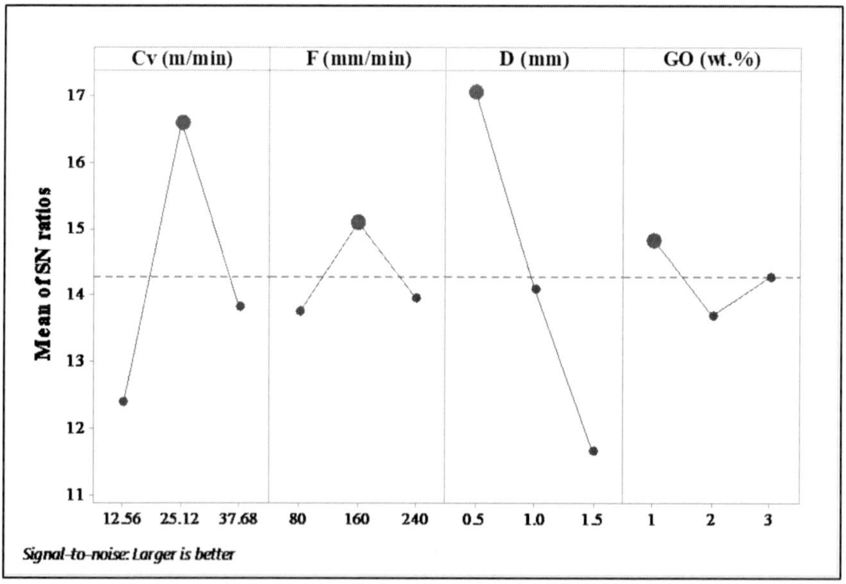

Figure 5.17. Optimal setting by CoCoSo-PCA for milling process (J. Kumar & Verma, 2021b).

The outcome was determined by adhering to the optimal parameter setting and its selected levels during the validation test. Table 5.5 illustrates Exp. 6, suggesting that the orthogonal array's most suitable machining performance is achieved with the optimal process parameter configuration of S2-F3-D1-GO2. Using the hybrid technique, the optimal setting of process parameters (S2-F2-D1-GO1) is illustrated in Figure 5.17, which displays the S / N ratio plot. The confirmatory test results achieved a preferred solution value of 9.35. It has been determined that the optimal parametric setting substantially enhanced the milling characteristics' desired quality and productivity. A decrease in Ra from 0.78 μm to 0.75 μm, an increase in the MRR from 15.97 mm3/min to 16.34 mm3/min, and a decrease in the cutting force from 5.60 N to 4.50 N are all visible. Surface roughness is reduced by 3.84%, cutting force is reduced by 19.64%, and MRR is improved by 2.26%. The 9.62% improvement in preferred solution values is extremely desirable for the cost-effective production of polymer components.

The correlation response, which is typically overlooked in conventional optimization modules, is addressed by the hybrid optimization module employed in this paper. Principal components (PCs) are uncorrelated indicators of machining performance, such as correlated responses that become independent. The CoCoSo module is a powerful statistical tool for resolving important problems in MCDM. According to state-of-the-art specifications, many modules and formulas for aggregating multiple features exist. This study investigates an important philosophy for combining several contradictory machining replies. Higher practical viability and an efficient method of resolving challenging optimization problems are offered by the CoCoSo-PCA coupling concept.

Conclusion

During the drilling and milling of CFRP epoxy nanocomposites modified with graphene oxide, the present study examines experimental investigation, statistically obtained observations, and multi-criteria optimization. An optimal condition for cost-effective machining responses with conflicting characteristics has been calculated by optimizing the following process constraints: feed rate (F), spindle speed (S)/cutting speed (S), depth of cut (D), and weight percent of graphene oxide (GO). The optimal parametric conditions have been evaluated efficiently by the CoCoSo-PCA module. The research proves that the hybridization strategy is possible. Based on the outcome, we can conclude the following:

- The most significant factor is feed rate, followed by spindle speed and GO (2.17%), as evidenced by the ANOVA results for response thrust force (Th), torque (T), delamination factor (IN & OUT), surface roughness (Ra & Rz), and circularity error. The model has been adequately developed.
- ANOVA results for response MRR, surface roughness (R_a), cutting force (F_c), and delamination factor demonstrate that cutting speed is the most influential factor, followed by feed rate, GO (2.17%), and depth of cut with the adequately developed model.
- The traditional optimization approach cannot accommodate the transition of the conflicting machining performances into a single objective function Ki.
- The purpose of PCA was to assign weights to machining responses, as decision-makers in most studies presume that all reactions are equally weighted. This assumption results in ambiguity, inaccuracy, and error in the results.
- The validation test confirms that the optimum parametric combination through CoCoSo-PCA is effective, as evidenced by the satisfactory accord with the actual values.
- For online and offline quality and productivity concerns, the hybridization methodology may be recommended for other manufacturing processes and multi-criteria optimization case studies.

This research emphasizes the effectiveness of a resilient hybridization strategy in machining LPNC. In the future, researchers can look into TWR, tool material, chip formation, material removal, and how heat acts., and health and safety hazards in polymer nanocomposites' machinability phases. The suggested module is a feature-rich optimization tool that may be adjusted for additional conventional and unconventional machining procedures. To comprehend more concerning the machining environment of LPNC, more work can be done in the future, such as choosing different factors like a different coolant, a different tip angle, a different cutter width, and a different material coated on the tool.

Present research work effectively addresses the development of GO/CF nanocomposites through a cost-effective fabrication method. After the development phase, the mechanical characterization shows the feasible result for the multifunctional component's fabrication. The machinability and machining performance optimization finding illustrates the proposed material's application potential and provides various ways to control quality and productivity concerns. Still, there is a vast scope for future work to understand better the proposed material manufacturing in society and trade interest. The proposed optimization modules in this series can be commended to other manufacturing methods such as turning, trimming, micromachining-based operations, etc. Manufacturing and academia may study, grow, and commercialize graphene oxide due to its outstanding properties. The issues of health and safety, mechanics of material removal, tool design, temperature effect, etc., are also emerging issues for researchers and academics.

The next chapter provides detailed information about the application potential and an evaluation of the proposed claims made in this book. For this, the industry-institute interface mode is used for the development of the product/prototype. An international organization for standardization (ISO: 9001:2015) certified industry collaborates to develop the end products and their probable functions as an industry partner. In this chapter, an effort has been rendered to convert the proposed laminated polymer nanocomposite into the probable end-product form. An attempt has been made to transform the academic laboratory-developed product to understand the industrial need for end-product applications.

References

Abhishek, K., Datta, S., & Mahapatra, S. S. (2014). Optimization of thrust, torque, entry, and exist delamination factor during drilling of CFRP composites. *International Journal of Advanced Manufacturing Technology*, *76*(1–4), 401–416.

Abhishek, K., Datta, S., & Mahapatra, S. S. (2016). Multi-objective optimization in drilling of CFRP (polyester) composites: Application of a fuzzy embedded harmony search (HS) algorithm. *Measurement: Journal of the International Measurement Confederation*, *77*, 222–239.

Ajith Arul Daniel, S., Pugazhenthi, R., Kumar, R., & Vijayananth, S. (2019). Multi objective prediction and optimization of control parameters in the milling of aluminium hybrid metal matrix composites using ANN and Taguchi -grey relational analysis. *Defence Technology*, *15*(4), 545–556.

Ali, R. A., Mia, M., Khan, A. M., Chen, W., Gupta, M. K., & Pruncu, C. I. (2019). Multi-response optimization of face milling performance considering tool path strategies in machining of Al-2024. *Materials*, *12*(7), 1–19.

Altintas, Y., Melbourne, E. A. U., & Berkeley, D. D. U. (1998). Present Situation and Future Trends in Modeling of Machining Operations Progress Report of the CIRP Working Group 'Modeling of Machining Operations.' *Cirp AnnalsManufacturing Technology*, *47*(2), 587–626.

An, Q., Cai, X., Xu, J., & Chen, M. (2014). Experimental investigation on drilling of high strength T800S / 250F CFRP with twist and dagger drill bits. *Int. J. Abrasive Technology*, *6*(3), 183–196.

Asghar, A., Raman, A. A. A., & Daud, W. M. A. W. (2014). A Comparison of Central Composite Design and Taguchi Method for Optimizing Fenton Process. *Scientific World Journal*, *2014*, 869120 (1-14).

Azmi, A. I., Lin, R. J. T., & Bhattacharyya, D. (2013). Machinability study of glass fibre-reinforced polymer composites during end milling. *International Journal of Advanced Manufacturing Technology*, *64*(1–4), 247–261.

Baraheni, M., & Amini, S. (2019). Comprehensive optimization of process parameters in rotary ultrasonic drilling of CFRP aimed at minimizing delamination. *International Journal of Lightweight Materials and Manufacture*, *2*(4), 379–387.

Bhattacharya, A., Das, S., Majumder, P., & Batish, A. (2009). Estimating the effect of cutting parameters on surface finish and power consumption during high speed machining of AISI 1045 steel using Taguchi design and ANOVA. *Production Engineering*, *3*(1), 31–40.

Bhopale, N. N., Pawade, R. S., & Joshi, S. S. (2017). Surface quality analysis in ball end milling of Inconel 718 cantilevers by response surface methodology. *Proceedings of the Institution of Mechanical Engineers, Part B: Journal of Engineering Manufacture*, *231*(4), 628–640.

Bielawski, R., Kowalik, M., Suprynowicz, K., Rzadkowski, W., & Pyrzanowski, P. (2017). Experimental study on the riveted joints in glass fibre reinforced plastics (GFRP). *Archive of Mechanical Engineering*, *64*(3), 301–313.

Coleman, J. N., Khan, U., Blau, W. J., & Gun'ko, Y. K. (2006). Small but strong: A review of the mechanical properties of carbon nanotube-polymer composites. *Carbon*, *44*(9), 1624–1652. https://doi.org/10.1016/j.carbon.2006.02.038.

Das, B., Roy, S., Rai, R. N., Saha, S. C., & Majumder, P. (2016). Effect of in-situ processing parameters on microstructure and mechanical properties of TiC particulate reinforced Al–4.5Cu alloy MMC fabricated by stir-casting technique – Optimization using grey based differential evolution algorithm. *Measurement: Journal of the International Measurement Confederation*, *93*, 397–408.

Das, M. K., Kumar, K., Barman, T. Kr., & Sahoo, P. (2014). Optimization of MRR and Surface Roughness in PAC of EN 31 Steel Using Weighted Principal Component Analysis. *Procedia Technology*, *14*, 211–218.

Datta, S., Bandyopadhyay, A., & Pal, P. K. (2008). Grey-based taguchi method for optimization of bead geometry in submerged arc bead-on-plate welding. *International Journal of Advanced Manufacturing Technology*, *39*(11–12), 1136–1143.

Datta, S., & Sankar Mahapatra, S. (2010). Use of desirability function and principal component analysis in grey-taguchi approach to solve correlated multi-response optimization in submerged arc welding. *Journal of Advanced Manufacturing Systems*, *9*(2), 117–128.

D.Lahane, Miss. Swati. (2012). Multi-response optimization of Wire-EDM process using principal component analysis. *IOSR Journal of Engineering*, *02*(08), 38–47.

Enemuoh, E. U., El-Gizawy, A. S., & Chukwujekwu Okafor, A. (2001). An approach for development of damage-free drilling of carbon fiber reinforced thermosets. *International Journal of Machine Tools and Manufacture*, *41*(12), 1795–1814.

Fu, R., Jia, Z., Wang, F., Jin, Y., Sun, D., Yang, L., & Cheng, D. (2018). Drill-exit temperature characteristics in drilling of UD and MD CFRP composites based on infrared thermography. *International Journal of Machine Tools and Manufacture*, *135*, 24–37. https://doi.org/10.1016/j.ijmachtools.2018.08.002 MTM.

Gaitonde, V. N., Karnik, S. R., Rubio, J. C., Correia, A. E., Abrão, A. M., & Davim, J. P. (2008). Analysis of parametric influence on delamination in high-speed drilling of carbon fiber reinforced plastic composites. *Journal of Materials Processing Technology*, *203*(1–3), 431–438.

Geier, N., & Szalay, T. (2017). Optimisation of process parameters for the orbital and conventional drilling of uni-directional carbon fibre-reinforced polymers (UD-CFRP). *Measurement: Journal of the International Measurement Confederation*, *110*, 319–334.

Gok, A. (2015). A new approach to minimization of the surface roughness and cutting force via fuzzy TOPSIS, multi-objective grey design and RSA. *Measurement: Journal of the International Measurement Confederation*, *70*, 100–109.

Gong, Y., & Li, P. (2019). Analysis of tool wear performance and surface quality in post milling of additive manufactured 316L stainless steel. *Journal of Mechanical Science and Technology*, *33*(5), 2387–2395.

Gopal, P. M. M., Prakash, K. S., & Soorya Prakash, K. (2017). Minimization of cutting force, temperature and surface roughness through GRA, TOPSIS and Taguchi techniques in end milling of Mg hybrid MMC. *Measurement: Journal of the International Measurement Confederation*, *116*, 178–192.

Harii Krishna Rao, G., & Ansari, M. N. M. (2013). Effect of cutting parameters on the surface roughness of MWCNT reinforced epoxy composite using end-milling process. *International Journal of Science and Research*, *2*(11), 02013372: (51-55).

Heidary, H., Karimi, N. Z., Minak, G., Zarif, N., Minak, G., Karimi, N. Z., & Minak, G. (2018). Investigation on delamination and flexural properties in drilling of carbon nanotube/polymer composites. *Composite Structures*, *201*, 112–120. https://doi.org/10.1016/j.compstruct.2018.06.041.

Hou, T. H., Su, C. H., & Liu, W. L. (2007). Parameters optimization of a nano-particle wet milling process using the Taguchi method, response surface method and genetic algorithm. *Powder Technology*, *173*(3), 153–162.

Islam, M. N., & Pramanik, A. (2016). Comparison of Design of Experiments via Traditional and Taguchi Method. *Journal of Advanced Manufacturing Systems*, *15*(3), 151–160.

Jenarthanan, M. P., & Jeyapaul, R. (2018). Optimisation of machining parameters on milling of GFRP composites by desirability function analysis using Taguchi method. *International Journal of Engineering, Science and Technology*, *5*(4), 22.

Jin, H., Tian, L., Bing, W., Zhao, J., & Ren, L. (2021). Toward the Application of Graphene for Combating Marine Biofouling. *Advanced Sustainable Systems*, *5*(1), 1–16. https://doi.org/10.1002/adsu.202000076.

Kalidass, S., & Palanisamy, P. (2014). Experimental investigation on the effect of tool geometry and cutting conditions using tool wear prediction model for end milling process. *Journal of Advanced Manufacturing Systems*, *13*(1), 41–54.

Karaşan, A., & Bolturk, E. (2019). Solid Waste Disposal Site Selection by Using Neutrosophic Combined Compromise Solution Method. *Proceedings of the 11th Conference of the European Society for Fuzzy Logic and Technology (EUSFLAT 2019)*, *1*, 416–422.

Karl Pearson F. R. S. (1901). On lines and planes of closest fit to systems of points in space. *The London, Edinburgh, and Dublin Philosophical Magazine and Journal of Science*, *2*(11), 559–572.

Kaybal, H. B., Ünüvar, A., Koyunbakan, M., & Avcı, A. (2019). A novelty optimization approach for drilling of CFRP nanocomposite laminates. *International Journal of Advanced Manufacturing Technology*, *100*(9–12), 2995–3012.

Khan, B., Adeleye, A. S., Burgess, R. M., Russo, S. M., & Ho, K. T. (2019). Effects of graphene oxide nanomaterial exposures on the marine bivalve, Crassostrea virginica. *Aquatic Toxicology*, *216*, 105297.

Kharwar, P. K., & Verma, R. K. (2019). Grey embedded in artificial neural network (ANN) based on hybrid optimization approach in machining of GFRP epoxy composites. *FME Transactions*, *47*(3), 641–648.

Kharwar, P. K., & Verma, R. K. (2020). Machining performance optimization in drilling of multiwall carbon nano tube/epoxy nanocomposites using GRA-PCA hybrid approach. *Measurement: Journal of the International Measurement Confederation*, *158*, 107701.

Kharwar, P. K., Verma, R. K., & Singh, A. (2020). Neural network modeling and combined compromise solution (CoCoSo) method for optimization of drilling performances in polymer nanocomposites. *Journal of Thermoplastic Composite Materials*, 1–28.

Kilickap, E. (2010). Optimization of cutting parameters on delamination based on Taguchi method during drilling of GFRP composite. *Expert Systems with Applications*, *37*(8), 6116–6122.

Köklü, U., Mayda, M., Morkavuk, S., Avci, A., & Demir, O. (2019). Optimization and prediction of thrust force, vibration and delamination in drilling of functionally graded composite using Taguchi, ANOVA and ANN analysis. *Materials Research Express*, *6*(8), 085335 (1-11).

Krishnamoorthy, A., Rajendra Boopathy, S., Palanikumar, K., & Paulo Davim, J. (2012). Application of grey fuzzy logic for the optimization of drilling parameters for CFRP composites with multiple performance characteristics. *Measurement: Journal of the International Measurement Confederation*, *45*(5), 1286–1296.

Krishnaraj, V., Prabukarthi, A., Ramanathan, A., Elanghovan, N., Kumar, M. S., Zitoune, R., & Davim, J. P. (2012). Optimization of machining parameters at high speed drilling of carbon fiber reinforced plastic (CFRP) laminates. *Composites Part B: Engineering*, *43*(4), 1791–1799.

Kulkarni, G. S., Shivashankar, G. S., Suresh, R., & Siddeshkumar, N. G. (2020). Optimization of Drilling Parameters of GFRP with Liquid Silicone Rubber and Fine Silica Powder by Taguchi Approach. *Silicon*, *12*(7), 1651–1666.

Kumar, D., & Singh, K. K. (2019). Investigation of delamination and surface quality of machined holes in drilling of multiwalled carbon nanotube doped epoxy/carbon fiber reinforced polymer nanocomposite. *Proceedings of the Institution of Mechanical Engineers, Part L: Journal of Materials: Design and Applications*, *233*(4), 647–663.

Kumar, J., & Kumar Verma, R. (2021). A novel methodology of Combined Compromise Solution and Principal Component Analysis (CoCoSo-PCA) for machinability investigation of graphene nanocomposites. *CIRP Journal of Manufacturing Science and Technology*, *33*, 143–157. https://doi.org/https://doi.org/10.1016/j.cirpj.2021.03.007.

Kumar, J., & Verma, R. K. (2021a). Experimental investigation for machinability aspects of graphene oxide/carbon fiber reinforced polymer nanocomposites and predictive modeling using hybrid approach. *Defence Technology*, *17*(5), 1671–1686. https://doi.org/10.1016/j.dt.2020.09.009.

Kumar, J., & Verma, R. K. (2021b). Multiple response optimization in machining (milling) of graphene oxide-doped epoxy/cfrp composite using CoCoSo-PCA: Anovel hybridization approach. *Journal of Advanced Manufacturing Systems*, *20*(2), 423–446. https://doi.org/10.1142/S0219686721500207.

Kumar, S., & Singh, R. (2019). Optimization of process parameters of metal inert gas welding with preheating on AISI 1018 mild steel using grey based Taguchi method. *Measurement: Journal of the International Measurement Confederation*, *148*, 106924.

Kyratsis, P., Markopoulos, A. P., Efkolidis, N., Maliagkas, V., & Kakoulis, K. (2018). Prediction of thrust force and cutting torque in drilling based on the response surface methodology. *Machines*, *6*(2), 24 (1-12).

Latha, B., & Senthilkumar, V. S. (2009). Analysis of thrust force in drilling glass fiber-reinforced plastic composites using fuzzy logic. *Materials and Manufacturing Processes*, *24*(4), 509–516.

Liu, Y. T., Chang, W. C., & Yamagata, Y. (2010). A study on optimal compensation cutting for an aspheric surface using the Taguchi method. *CIRP Journal of Manufacturing Science and Technology*, *3*(1), 40–48.

Mi, X., & Liao, H. (2020). Renewable energy investments by a combined compromise solution method with stochastic information. *Journal of Cleaner Production*, *276*, 123351 (1-13). https://doi.org/10.1016/ j.jclepro.2020.123351.

Mohan, N. S., Ramachandra, A., & Kulkarni, S. M. (2005). Influence of process parameters on cutting force and torque during drilling of glass-fiber polyester reinforced composites. *Composite Structures*, *71*(3–4), 407–413. https://doi.org/10.1016/j.compstruct.2005.09.039.

Mourtzis, D., Doukas, M., & Bernidaki, D. (2014). Simulation in manufacturing: Review and challenges. *Procedia CIRP*, *25*, 213–229.

Naveen Sait, A., Aravindan, S., & Noorul Haq, A. (2009). Optimisation of machining parameters of glass-fibre-reinforced plastic (GFRP) pipes by desirability function analysis using Taguchi technique. *International Journal of Advanced Manufacturing Technology*, *43*(5–6), 581–589. https://doi.org/10.1007/s00170-008-1731-y.

Palanikumar, K. (2011). Experimental investigation and optimisation in drilling of GFRP composites. *Measurement: Journal of the International Measurement Confederation*, *44*(10), 2138–2148.

Palanikumar, K., Karunamoorthy, L., & Karthikeyan, R. (2006). Multiple performance optimization of machining parameters on the machining of GFRP composites using carbide (K10) tool. *Materials and Manufacturing Processes*, *21*(8), 846–852.

Panchagnula, K. K., & Palaniyandi, K. (2018). Drilling on fiber reinforced polymer/nanopolymer composite laminates: A review. *Journal of Materials Research and Technology*, *7*(2), 180–189.

Papageorgiou, D. G., Liu, M., Li, Z., Vallés, C., Young, R. J., & Kinloch, I. A. (2019). Hybrid poly(ether ether ketone) composites reinforced with a combination of carbon fibres and graphene nanoplatelets. *Composites Science and Technology*, *175*, 60–68.

Parida, A. K., & Routara, B. C. (2014). Multiresponse Optimization of Process Parameters in Turning of GFRP Using TOPSIS Method. *International Scholarly Research Notices*, *2014*, 1–10.

Parida, A. kumar, Das, R., Sahoo, A. K., & Routara, B. C. (2014). Optimization of Cutting Parameters for Surface Roughness in Machining of gfrp Composites with Graphite/fly Ash Filler. *Procedia Materials Science*, *6*, 1533–1538.

Patel, K., Batish, A., & Bhattacharya, A. (2009). Optimization of surface roughness in an end-milling operation using nested experimental design. *Production Engineering*, *3*(4–5), 361–373.

Premnath, A., Alwarsamy, T., & Sugapriya, K. (2014). A comparative analysis of tool wear prediction using response surface methodology and artificial neural networks. *Australian Journal of Mechanical Engineering*, *12*(1), 38–48.

Qiu, X., Li, P., Li, C., Niu, Q., Chen, A., Ouyang, P., & Ko, T. J. (2018). Study on chisel edge drilling behavior and step drill structure on delamination in drilling CFRP. *Composite Structures*, *203*, 404–413.

Ragunath, S., Velmurugan, C., & Kannan, T. (2017). Optimization of drilling delamination behavior of GFRP/clay nano-composites using RSM and GRA methods. *Fibers and Polymers*, *18*(12), 2400–2409.

Rajesh Mathivanan, N., Mahesh, B. S., & Anup Shetty, H. (2016). An experimental investigation on the process parameters influencing machining forces during milling of carbon and glass fiber laminates. *Measurement: Journal of the International Measurement Confederation*, *91*, 39–45.

Rajkumar, D., Ranjithkumar, P., & Narayanan, C. S. (2017). Optimization of machining parameters on microdrilling of CFRP composites by taguchi based desirability function analysis. *Indian Journal of Engineering and Materials Sciences*, *24*(5), 331–338.

Raymond H. Myers, Douglas C. Montgomery, C. M. A.-C. (2016). Response Surface Methodology. In *Response Surface Methodology: Process and Product Optimization Using Designed Experiments*.

Routara, B. C., Mohanty, S. D., Datta, S., Bandyopadhyay, A., & Mahapatra, S. S. (2010). Combined quality loss (CQL) concept in WPCA-based Taguchi philosophy for optimization of multiple surface quality characteristics of UNS C34000 brass in cylindrical grinding. *International Journal of Advanced Manufacturing Technology*, *51*(1–4), 135–143.

Sahoo, P., Pratap, A., & Bandyopadhyay, A. (2017). Modeling and optimization of surface roughness and tool vibration in CNC turning of aluminum alloy using hybrid RSM-

WPCA methodology. *International Journal of Industrial Engineering Computations*, *8*(3), 385–398.

Sampath Kumar, T., Balasivanandha Prabu, S., & Sorna Kumar, T. (2017). Comparative Evaluation of Performances of TiAlN-, AlCrN- and AlCrN/TiAlN-Coated Carbide Cutting Tools and Uncoated Carbide Cutting Tools on Turning EN24 Alloy Steel. *Journal of Advanced Manufacturing Systems*, *16*(3), 237–261.

Sardiñas, R. Q., Reis, P., & Davim, J. P. (2006). Multi-objective optimization of cutting parameters for drilling laminate composite materials by using genetic algorithms. *Composites Science and Technology*, *66*(15), 3083–3088.

Sarikaya, M., & Güllü, A. (2014). Taguchi design and response surface methodology based analysis of machining parameters in CNC turning under MQL. *Journal of Cleaner Production*, *65*, 604–616.

Selvarajan, L., Narayanan, C. S., & Jeyapaul, R. (2015). Optimization of EDM Hole Drilling Parameters in Machining of MoSi2-SiC Intermetallic/Composites for Improving Geometrical Tolerances. *Journal of Advanced Manufacturing Systems*, *14*(4), 259–272.

Senthilkumar, N., Tamizharasan, T., & Gobikannan, S. (2014). Application of Response Surface Methodology and Firefly Algorithm for Optimizing Multiple Responses in Turning AISI 1045 Steel. *Arabian Journal for Science and Engineering*, *39*(11), 8015–8030. https://doi.org/10.1007/s13369-014-1320-3.

Shunmugesh, K., & Panneerselvam, K. (2016). Optimization of process parameters in micro-drilling of Carbon Fiber Reinforced Polymer (CFRP) using taguchi and grey relational analysis. *Polymers and Polymer Composites*, *24*(7), 499–506.

Shunmugesh, K., & Panneerselvam, K. (2017). Optimization of machining process parameters in drilling of CFRP using multi-objective Taguchi technique, to PSIS and RSA techniques. *Polymers and Polymer Composites*, *25*(3), 185–192.

Singaravel, B., & Selvaraj, T. (2016). Application of Desirability Function Analysis and Utility Concept for Selection of Optimum Cutting Parameters in Turning Operation. *Journal of Advanced Manufacturing Systems*, *15*(1), 1–11.

Sonkar, V., Abhishek, K., Datta, S., & Mahapatra, S. S. (2014). Multi-objective Optimization in Drilling of GFRP Composites: A Degree of Similarity Approach. *Procedia Materials Science*, *6*, 538–543.

Sorrentino, L., Turchetta, S., & Bellini, C. (2017). In process monitoring of cutting temperature during the drilling of FRP laminate. *Composite Structures*, *168*, 549–561.

Sridharan, V., Raja, T., & Muthukrishnan, N. (2016). Study of the Effect of Matrix, Fibre Treatment and Graphene on Delamination by Drilling Jute/Epoxy Nanohybrid Composite. *Arabian Journal for Science and Engineering*, *41*(5), 1883–1894. https://doi.org/10.1007/s13369-015-2005-2.

Tam, S. C., Yeo, C. Y., Jana, S., Lau, M. W. S., Lim, L. E. N., Yang, L. J., & Noor, Y. M. (1993). Optimization of laser deep-hole drilling of Inconel 718 using the Taguchi method. *Journal of Materials Processing Tech.*, *37*(1–4), 741–757.

Thakur, R. K., Singh, K. K., & Kumar, K. (2020). Investigation of milling characteristics in graphene-embedded epoxy/carbon fibre reinforced composite. *Materials Today: Proceedings*, *33*, 5643–5648.

Thirukkumaran, K., Menaka, M., Mukhopadhyay, C. K., & Venkatraman, B. (2020). A Study on Temperature Rise, Tool Wear, and Surface Roughness During Drilling of Al–5%SiC Composite. *Arabian Journal for Science and Engineering*, *45*(7), 5407–5419. https://doi.org/10.1007/s13369-020-04427-4.

Tsao, C. C. (2008). Thrust force and delamination of core-saw drill during drilling of carbon fiber reinforced plastics (CFRP). *International Journal of Advanced Manufacturing Technology*, *37*(1–2), 23–28.

Tsao, C. C. (2009). Grey-Taguchi method to optimize the milling parameters of aluminum alloy. *International Journal of Advanced Manufacturing Technology*, *40*(1–2), 41–48. https://doi.org/10.1007/s00170-007-1314-3.

Tsao, C. C., & Hocheng, H. (2008). Evaluation of thrust force and surface roughness in drilling composite material using Taguchi analysis and neural network. *Journal of Materials Processing Technology*, *203*(1–3), 342–348.

Uzun, G. (2019). Analysis of grey relational method of the effects on machinability performance on austempered vermicular graphite cast irons. *Measurement: Journal of the International Measurement Confederation*, *142*, 122–130.

Vankanti, V. K., & Ganta, V. (2014). Optimization of process parameters in drilling of GFRP composite using Taguchi method. *Journal of Materials Research and Technology*, *3*(1), 35–41.

Vasudevan, H., Deshpande, N., Rajguru, R., & Mane, S. (2014). Grey fuzzy multiobjective optimization of process parameters for CNC turning of GFRP/Epoxy Composites. *Procedia Engineering*, *97*, 85–94.

Vasudevan, H., Rajguru, R., Tank, K., & Shetty, N. (2018). Optimization of Multi-Performance Characteristics in the Turning of GFRP(E) Composites using Principle Component Analysis combined with Grey Relational Analysis. *Materials Today: Proceedings*, *5*(2), 5955–5967.

Wang, X., Wang, L. J., & Tao, J. P. (2004). Investigation on thrust in vibration drilling of fiber-reinforced plastics. *Journal of Materials Processing Technology*, *148*(2), 239–244.

Weglowski, M. S. (2014). Experimental study and response surface methodology for investigation of FSP process. *Archive of Mechanical Engineering*, *LXI*(4), 539–552.

Wen, Z., Liao, H., Kazimieras Zavadskas, E., & Al-Barakati, A. (2019). Selection third-party logistics service providers in supply chain finance by a hesitant fuzzy linguistic combined compromise solution method. *Economic Research-Ekonomska Istrazivanja*, *32*(1), 4033–4058.

Xu, J., An, Q., Cai, X., & Chen, M. (2013). Drilling machinability evaluation on new developed high-strength T800S/250F CFRP laminates. *International Journal of Precision Engineering and Manufacturing*, *14*(10), 1687–1696.

Xu, J., Ji, M., Chen, M., & Ren, F. (2019a). Investigation of minimum quantity lubrication effects in drilling CFRP / Ti6Al4V stacks. *Materials and Manufacturing Processes*, *34*(12), 1401–1410.

Xu, J., Ji, M., Chen, M., & Ren, F. (2019b). Investigation of minimum quantity lubrication effects in drilling CFRP / Ti6Al4V stacks. *Materials and Manufacturing Processes*, *34*(12), 1401–1410.

Xu, J., Ji, M., Paulo Davim, J., Chen, M., El Mansori, M., & Krishnaraj, V. (2020). Comparative study of minimum quantity lubrication and dry drilling of CFRP/titanium stacks using TiAlN and diamond coated drills. *Composite Structures*, *234*, 111727.

Xu, J., Li, C., Mi, S., An, Q., & Chen, M. (2018). Study of drilling-induced defects for CFRP composites using new criteria. *Composite Structures*, *201*, 1076–1087.

Yan, J., & Li, L. (2013). Multi-objective optimization of milling parameters-the trade-offs between energy, production rate and cutting quality. *Journal of Cleaner Production*, *52*, 462–471.

Yang, H., Luo, R., Han, S., & Li, M. (2010). Effect of the ratio of graphite/pitch coke on the mechanical and tribological properties of copper-carbon composites. *Wear*, *268*(11–12), 1337–1341.

Yazdani, M., Zarate, P., Kazimieras Zavadskas, E., & Turskis, Z. (2019). A combined compromise solution (CoCoSo) method for multi-criteria decision-making problems. *Management Decision*, *57*(9), 2501–2519. https://doi.org/10.1108/MD-05-2017-0458.

Chapter 6

Evaluation of Machining Characteristics of Laminated Polymer Nanocomposites: Unconventional Machining Interface

The unconventional machining process is progressively utilized to provide a high-quality finished product of laminated polymer composites (LPC) and laminated polymer nanocomposite (LPNC). Unconventional machining technologies reduced machining time and errors while increasing the quality of the machined surface in the laminated composite. Technology like temperature, chemical, pressure, frequency vibration, and phase transformation mechanisms remove material in nontraditional processing. On several significant technologies, abrasive waterjet machining (AWJM), electric discharge machining (EDM)/wire-EDM (WEDM)/micro-EDM (μEDM), and laser beam machining (LBM) are in development. However, only a few processes have been utilized to machine LPNC.

6.1. Introduction to the Unconventional Machining Process

In contrast to the process of machining metal, composite material processing has different results. It is challenging to minimize machining defects due to its low glass transition temperature, high fiber reinforcement abrasion, and heterogeneous and anisotropic properties. These features indicate that LPC still experiences many various types of defects. Delamination (Fd), fiber pull-out, kerf taper (Kt), surface roughness (Ra), and fiber-matrix debonding are typical modes of failure after machining (Natarajan et al., 2020). Consequential rework and waste were generated due to defects in the composite parts. Additional inspections, decreased output, and decreased customer satisfaction impose financial and time expenses on an organization. The objective of the research societies was to overcome the limitations of conventional processes and mitigate these issues by focusing on the study of various non-conventional machining techniques for LPC (Jain, 2008). Non-conventional machining processes utilize different energy sources, including mechanical, thermal, electrochemical, or various energy types, to produce the intended final geometry. Illustrated in Figure 6.1 are prevalent classifications of unconventional machining procedures.

Mechanical, thermal, and electrochemical machining are the three unconventional methods introduced. The erosion mechanism in a mechanical type uses hydraulic pressure and high-velocity abrasive particles to remove material from the workpiece. Mechanical machining includes processes such as ultrasonic machining (USM) and AWJM. The thermal and electrical forms use intensified light or electricity to evaporate the undesirable material of the workpiece. It includes processes such as EDM, WEDM, and LBM. Electrochemical machining (ECM) is a process that uses a high-current electric field to dissolve material in an electrolyte solution using an ion displacement mechanism.

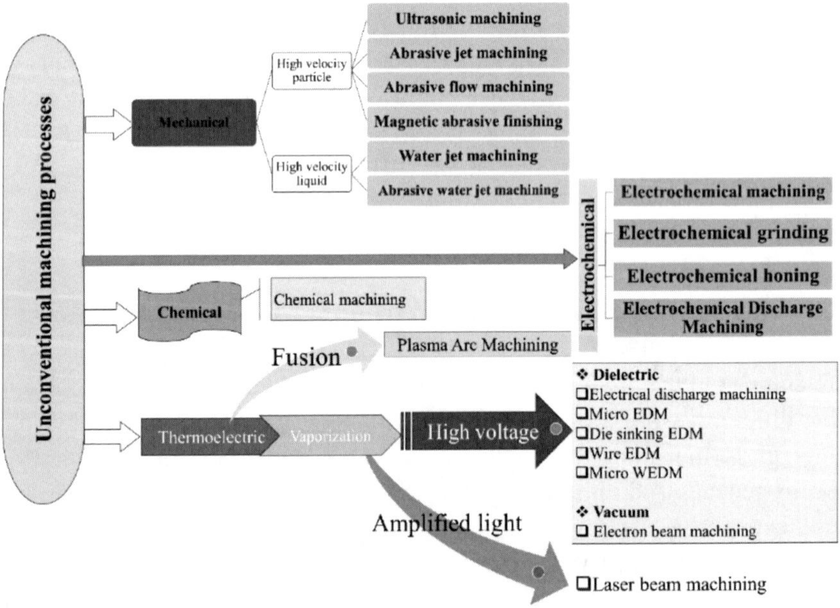

Figure 6.1. Classification of unconventional machining processes (Prasad & Chakraborty, 2018).

The best non-traditional machining techniques for LPC/LPNC processing include AWJM, LBM, and EDM. Machining LPC/LPNC involves a cutting mechanism, and this part gives an overview of the technologies employed in each operation.

6.2. Abrasive Water Jet Machining

With the AWJM method, you may achieve cold cutting precision while minimizing thermal stress. By using water to dissipate heat, the AWJM technology makes it possible to cut composites without compromising their properties. The specimen is exposed to a high-pressure water jet that contains fine-grained abrasive particles of aluminum oxide or garnet through a nozzle. An abrasive particle on the specimen surface causes erosion and material removal. An unconventional method of cutting, waterjet cutting uses a highly pressurized water stream to cut materials. Pure waterjet machining (WJM) and AWJM are the two main categories of waterjet cutting systems. WJM uses clean, high-pressure water to slice through rubber, plywood, and other soft materials. The AWJM can cut through tougher and more brittle materials such as advanced composites, glass, and construction materials by combining abrasive using waterjet mixes in the mixing chamber. Figure 6.2 (a) shows the AWJM process setup and Figure 6.2 (b) the kerf width of machining the laminated workpiece.

The system is designed to redirect the high-pressure clean water into a collimated jet and speed it up once it enters the orifice. The next step is to force the abrasive particles and air down using the suction pressure that the clean water jet produces. Introducing abrasive particles into the mixing chamber is time-dependent to avoid mixing tube blockages. A high-velocity water jet is combined in the mixing chamber with abrasive particles, like aluminum oxide and garnet, which flow from the abrasive input. These particles often show AWJM's erosive action. In the end, the abrasive particle and waterjet mixture is accelerated and then sent via a mixing tube that is normally 45 mm long and has an internal diameter of 1 mm. The jet nozzle is a pathway for the high-velocity exit jet stream, ultimately impacting the workpiece.

Choosing appropriate process parameters is the most critical step in making a high-quality product using AWJM for machining (Gopal et al., 2023; Mesalamy & Youssef, 2020). The strong relationship between these process parameters and the process is demonstrated in Figure 6.3. The AWJM method efficiency and geometric accuracy of LPC/LPNC were included in the output response, together with any defects or geometrical damage to the material. The machining performance in AWJM depends on machining parameters since this cold-working process causes heat-affected zones, reduced machining force, and microscopic thermal stress. The material is readily available during machining, free from the stresses and strains of the work process. The machining arrangement for the AWJM technique is illustrated in Figure 4.4.

6.2.1. AWJM Mechanism to Material Removal Rate (MRR)

The AWJM process involves using water as a medium to accelerate the abrasive particles onto the workpiece. The specimen's top surface is the first to be sliced by the high-momentum abrasive particles since they have greater energy. However, the formation of bent water jets occurs because the particles continually lose kinetic energy while the water jet is propagated. AWJM can accomplish two types of material removal processes: micro-mechanism and macro-mechanism. The low-energy stream is limited to removing the matrix between the fibers. To be more precise, the micro-mechanism of AWJM involves the rapid impact of abrasive particles on the machined sample.

The process of kerf generation is shown by macro-mechanism. The striated surface of the cut is created when the abrasive water jet penetrates to a specific depth during the cutting process. The fundamental step of abrasive waterjet cutting, involves the high impact impingement of solid abrasive particles to remove material. The material properties' ductile and brittle characteristics provide additional opportunities to distinguish between these processes. Plowing and cutting deformation are the two removal modes comprising the machining process for ductile materials. Brittle failure in an AWJM case would lead to the machining out of the undesirable material due to fracture initiation, propagation, and plastic deformation. The macro-mechanism of cutting is the development of a striated cut surface and the penetration depth of the abrasive water jet. The generated cutting front ends often have their geometry at the surface that is striated or wavecut. At the jet entrance (top), the machined surface is smoother than at the jet exit. The kerf wall roughness of FRP composite is divided into three parts: the damaged area, and the smooth and rough cutting area. These occurrences are caused by the abrasive water jet losing kinetic energy continuously while machining.

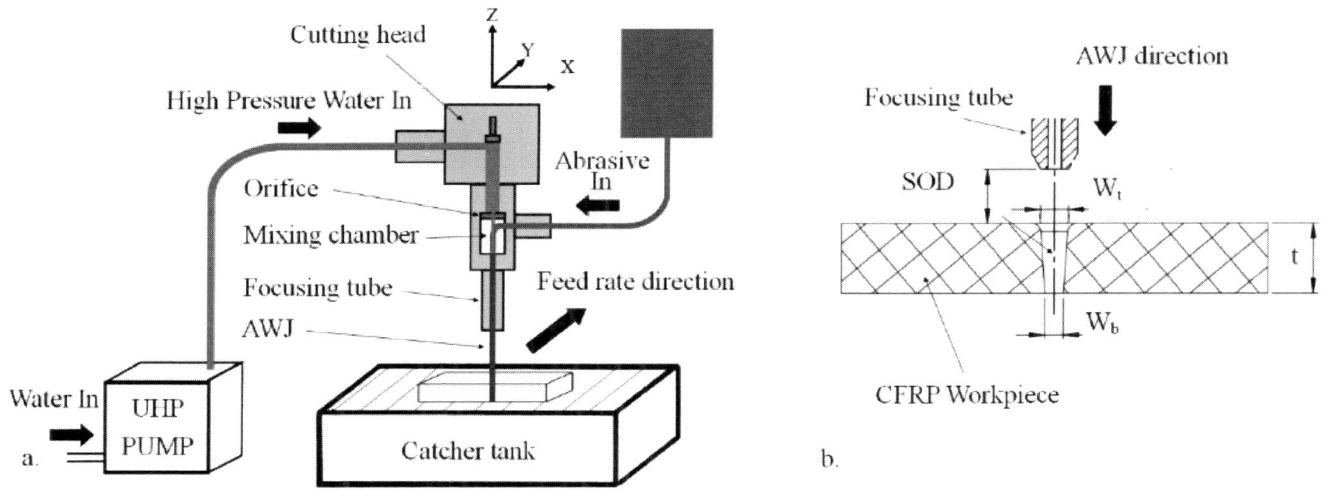

Figure 6.2. (a) AWJM setup (b) kerf geometry (Popan et al., 2023; H. Wang et al., 2023).

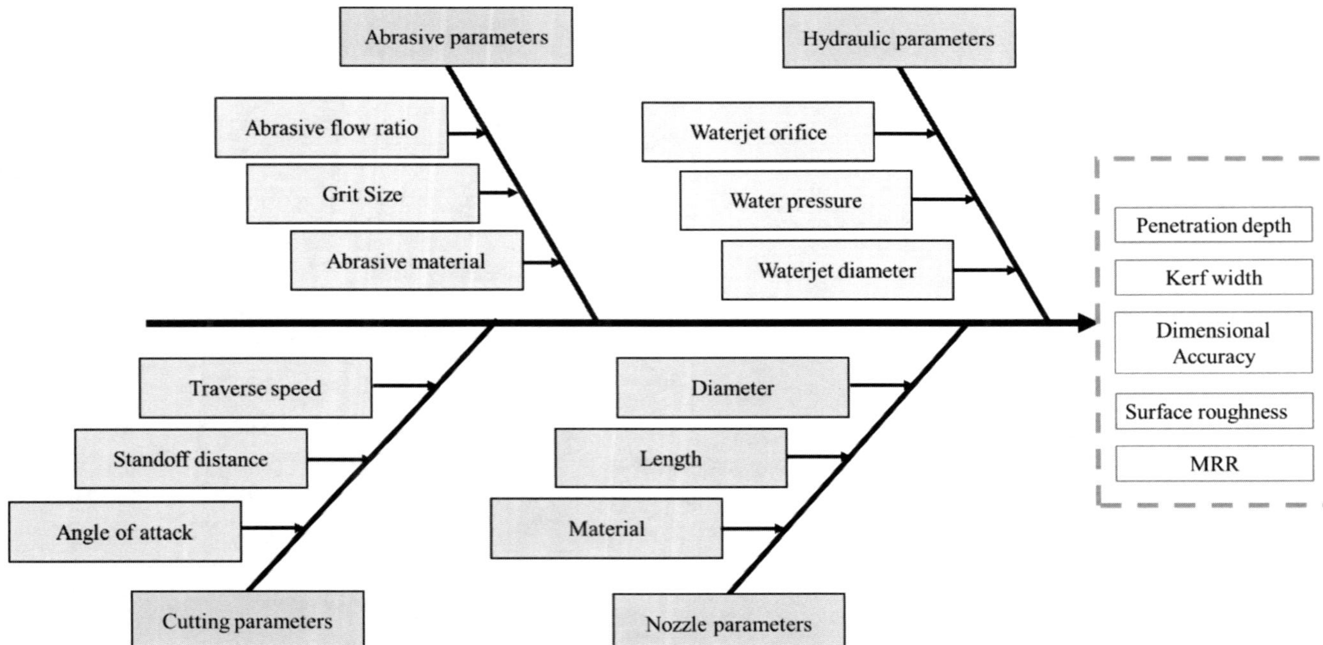

Figure 6.3. Fishbone diagram for the AWJM process of LPNC (Anu Kuttan et al., 2021; Kechagias et al., 2012).

Table 6.1. Delamination, Kerf tapper, and surface quality of LPC/LNPC

Ref.	Process Parameter	Effect	Observation	Validation Method
(Deepa et al., 2020)	C-N Bond Formation	Formation of C=C, C-F, and Halogen bonds after AWJ cutting	Structural Integrity: SEM analysis confirms the structural integrity and damage assessment of SRCs. Polymeric Changes: Structural changes post-cutting enhance mechanical properties, particularly after AWJ cutting.	Increased by 20% compared to conventional cut samples by analysis Taguchi optimized method
(Singh & Thakur, 2022)	MWCNT wt.%, JP, Tr, SoD	Ra, Kt, MRR, Fd (IN and OUT)	Systematic optimization of machining parameters leads to improved machining performance.	Optimized using Taguchi-based analysis with GRA and ANOVA
(Thakur et al., 2021)	JP (96 and 304 MPa), Tr (50, 100, and 150 mm/min), Nanoclay content (0.00, 1.25, and 2.50 wt.%)	CFRP with nano clay shows lower KT formation, Ra, and MRR than CFRP without nanoclay. A higher traverse rate (150 mm/min) results in greater Ra, KT	Higher JP (304 MPa) leads to lower Ra and Kt formation and improved MRR.	Taguchi-based statistical method for experimental design, ANOVA to assess the influence of input factors

6.2.2. Evaluation of AWJM Efficiency (Delamination, Kerf Width, Surface Roughness)

Vigneshwaran et al., (Vigneshwaran et al., 2018) report on the sisal fiber composites. The investigation found that the AWJM fiber-cutting mechanism differs entirely from traditional machining. Kalusuraman et al., (Kalusuraman et al., 2020) described the AWJM effect on luffa cylindrical fiber composites. Performance has been successfully reduced according to the experimental results. On composites made of filled hemp and Kevlar fibers, Fd was reduced when filler was added to the fiber mixture, as researched by Jani et al., (Jani et al., 2016). The cutting performance of fiber composites in AWJM is determined by machining factors such as stand-off distance (SOD), transverse speed, and cutting water pressure. The effectiveness of AWJM is measured by various output characteristics, including the rate at which material is removed, the shape and dimensions of the cut (known as kerf geometry), the occurrence of Fd (separation of layers), and the roughness of the resulting surface. The most frequent failures in fiber composites when using AWJM are Fd and the production of kerf geometry. These failures are undesirable because of the abrasive water jet's high-velocity impact and the fiber composites' properties. Cutting fiber composites in AWJM efficiently requires careful selection and optimization of machining parameters. The green polypropylene composite shows striation markings on the cut surface due to variations in water pressure. The water jet dynamic properties are primarily responsible for the considerable rise in striation markings near the cut surface. The SOD has a substantial impact on the fiber composite Kt and ratio. A higher Kt is the result of a high jet divergence, which occurs at high SOD. Based on their research, Jagadish et al., (Jagadish et al., 2016) determined that the two most important characteristics affecting the surface quality of green composites are the flow rate and abrasive particle size. According to their findings, AWJM is the best machining technique for cutting polymer composites that contain wood dust. More Fd occurred in the hybrid composite on AWJM due to the weak matrix bonding of flax and carbon fiber, and interlaminar Fd occurred on the cut surface due to the hydrophilic character of the natural fiber. Jagdish et al., (Jagadish et al., 2018) studied the action of AWJM on composites made of pine fibers. Machining at 100 MPa water pressure, 2 mm SOD, and 90 mesh abrasive particle size results in the lowest Ra of 0.13 mm. Kerf width ranged from 0.205 mm to 0.575 mm, while Ra values ranged from 3.141 mm to 6.538 mm for the composite of Kenaf and glass fiber. While the matrix region showed smooth cutting, the fiber region exhibited uneven cuts when Kalirasu et al., (Kalirasu S et al., 2016) studied the abrasive waterjet cutting process on jute fiber composite. When cutting with an abrasive water jet, the interfacial bonding properties of the matrix and reinforcement play a significant role in natural fiber composites. The banana fiber composites had vastly different surface qualities depending on the specific abrasive particles used. While performing chemical treatment, the Ra of the composite made of coconut sheath fibers is reduced by 12-22%. Through AWJM, Hutyrova (Hutyrová et al., 2016) performed a turning operation on a wood-plastic composite. Reduced transverse speed was associated with reduced Ra in the experiments. According to the research, the matrix and reinforcement mostly affect the machined composite Ra. A study on the cutting and machinability of LPC/LPNC was found by several researchers, as shown in Table 6.1.

6.3. EDM, WEDM, and Micro-EDM

Another possible alternative procedure for producing FRPs with great dimensional accuracy and complex shapes is EDM. In contrast to AWJM and LBM, EDM employs an electrode or wire to complete material removal procedures. Despite this, EDM can create precise components for materials that are difficult to machine or electrically conductive since the tool and the workpiece do not come into direct physical touch. There are two main categories of EDM: die-sinking EDM and wire EDM. The electrode forms vary, while the underlying principle is the same for all these machining kinds. By using triggered spark discharges and dielectric fluid, electrode machining, which is also sometimes called "spark erosion" machining, removes very small amounts of material between the tool and the workpiece. In addition to eliminating waste material from the sparking area, the dielectric fluid serves as a coolant and regulates the sparking distance between the tool and the workpiece.

In the first step, the dielectric fluid prevents current from flowing between the tool and the workpiece. The generation of an influential electrical field occurs to bridge the distance between the tool and the workpiece. A high-temperature spark eventually erodes or melts the surface of the workpiece after the dielectric fluid is penetrated by a transient spark that is generated as a result of the dielectric's breakdown. Carbon fiber is an EDM-machinable material that is electrically conductive, unlike aramid and glass fiber. Nevertheless, EDM machining has become more challenging with the introduction of semiconducting qualities into CFRPs composed of unique carbon fiber and polymer matrix characteristics. Choosing the right process parameters is crucial in achieving high-quality results following EDM machining. An array of processing parameters categorized by wire electrode, electrical, non-electrical, and workpiece parameters is shown in Figure 6.4. Due to their ability to regulate the power (p), the voltage (V), tool material, and pulse on-off duration were essential aspects in the EDM process of CFRPs. The feasibility of employing EDM to machine CFRPs was pioneered by experiments carried out by Lau et al., (Lau et al., 1990). The experimental findings demonstrated that, in comparison to graphite electrodes, copper electrodes result in superior tool wear (TWR) and surface polish. It was also mentioned that a low MRR would be produced by a high current density. A sluggish discharge rate is caused by an excess of melted matrix, which results from an excess of thermal energy in the workpiece. To minimize thermal damage to the machining surface and keep productivity at its best, it is crucial to correctly choose the process parameter range.

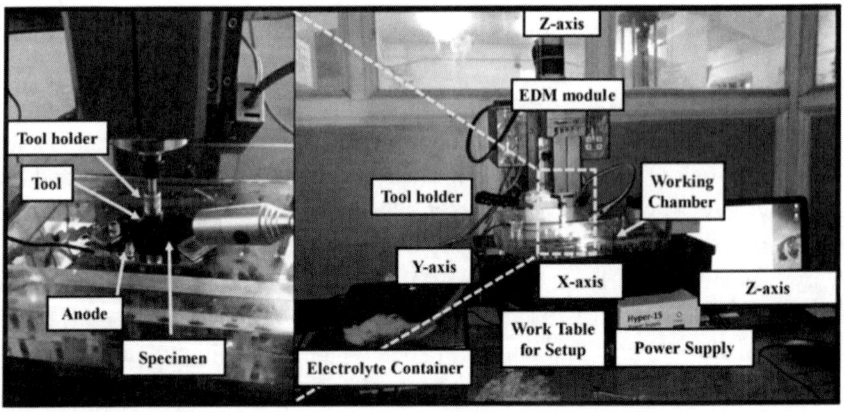

Figure 6.4. Setup of micro-EDM process (Vishwakarma et al., 2023).

6.3.1. EDM Mechanism for Material Removal Rate (MRR)

Material is removed through thermal erosion following a series of spark discharges that are created between the electrode and the workpiece. The gap is filled with a strong electric field as the electrode and workpiece receive a DC pulse (pulse on time). A small amount of dielectric breakdown and deionization causes the spark to be produced. A small amount of material from the electrode and workpiece melts and vaporizes when a high-temperature spark discharge is generated. The insulating properties of the dielectric are restored when new dielectric fluid moves into place across the gaps created by the temporary cessation of pulse flow (pulse off time). Because of their one-of-a-kind combination of highly conductive carbon fibers and a nonconductive matrix, CFRPs have unique features that have led to a lack of research into the mechanisms by which the EDM process causes them to have relatively low bulk electrical conductivity. A typical WEDM machine has a higher rate of wire breakage failure due to a highly localized concentration matrix. Recent research has developed and executed the preheating-assisted WEDM (PAWEDM) procedure to address this issue. Prior to the WEDM procedure, CFRPs had their mechanical performance enhanced using a sandwich-assisting electrode technique. The procedure was implemented to commence erosion, which could be done efficiently even if CFRP is semiconductive. Table 6.2 outlines preliminary work on the AWJM of PLC/PLNCs.

6.3.2. Evaluation of EDM Efficiency (MRR, Kerf Width, and Surface Roughness)

CLPC is a solid and lightweight composite material. Due to their superior modulus, strength, and low density, CLPC composites are widely used as metal replacements, such as steel and metallic alloys, and they have many applications in the aerospace, fiber optics, and automotive fields. Traditional mechanical machining procedures, such as milling and drilling, have some drawbacks, including Fd, splintering, burrs on the machined surface, and a short TWR. EDM is one of the unconventional machining technologies that is excellent for machining difficult-to-cut materials. It is one of the advanced machining processes that uses thermal energy to machine electrically conduct materials regardless of their hardness properties. Micro-EDM is discussed in detail as the most modern machining process. This section focuses on the complex material removal mechanism of CLPC due to the partially electrically conductive carbon fabric and the epoxy nonconductive nature. Different methods were applied to machining CLPC material by EDM, WEDM, and μEDM characteristics, such as the assisting electrode and reinforcing conductive particle methods. It also focused on high aspect ratio, hole dilation, analysis of machined surface, parametric influences, applications, and recent advancements for CLPC laminates.

The EDM drilling procedure of a unidirectional composite material was investigated by Jimenez et al., (Roldan-Jimenez et al., 2022) They were assessed from a macro geometrical perspective by looking at their surface quality, optimal diameter, and conicity. In addition, the wear of the electrode was evaluated through visual inspection. Because of the material anisotropy, there were noticeable variations in surface quality across the two orientations measured: parallel (0-180◦) and perpendicular (90-270◦). A minimum Ra value of 0.69 μm and 2.69 in the 90-270◦ orientation were observed for the measurements taken in the 0-180◦ orientations, indicating that the current and intensity values were substantial. Higher values of these factors are found to result in greater discharges, which in turn cause larger craters and worse surface quality. The stress variables and pulse time are directly related to the dimensional variation between the hole entry and exit. In an experimental study, Dutta et al., (Dutta et al., 2021) used sandwich-assisting electrodes to facilitate wire electrical-discharge machining (WEDM) to cut a thin carbon fiber-reinforced polymer (CFRP) composite plate. The analysis was used to examine whether the input parameters, input current, pulse on time, pulse off time, and voltage affected the cutting time and the machined surface morphology. They observed that the cutting time increased with an increase in pulse-off time and decreased with an increase in current.

Table 6.2. Investigation of delamination, Kerf tapper, and surface quality of LPC/LNPC

Ref.	Types of Machining	Process Parameter	Effect	Observation	Validation Method
(Ablyaz et al., 2021)	WEDM	Voltage, pulse duration	Cut-width (kerf)	Voltage and pulse duration significantly affect cut-width, Accuracy of PCM	Analysis of variance (ANOVA)
(Vishwakarma & Verma, 2021)	μEDM	Voltage (3 levels), Pulse Length (2 levels), GNP Wt.% (3 levels)	Higher MRR at optimal setting	Improved MRR and better micromachining quality due to GNP addition.	Taguchi orthogonal array and Sine Cosine Algorithm for MRR.

6.4. Laser Machining

Owing to its desirable qualities like a high strength-to-weight ratio, particular stiffness, and corrosion resistance, LPC is a vital engineering material utilized in various industries, including aviation, maritime, railways, energy, cars, military, and sports. Manufacturing components used in multiple sectors involve performing conventional machining processes such as drilling, milling, turning, and shaping. Their abrasiveness and anisotropic characteristics make machining these materials difficult. Traditional CFRP machining can lead to several issues, including TWR, matrix pull-out, fiber breakage, and Fd. Laser beam machining and AWJ are two of the most common unconventional machining processes. The strength of composite laminates is reduced during AWJ machining because of moisture absorption, abrasive particle trapping, and Fd. Ultrasonic machining has been utilized in place of WJM. Ultrasonic machining produces high-quality cuts, but it wears down tools quickly. As an alternative to traditional machining, laser processing of CFRP material was initially offered in the 1990s. This procedure ensured higher finishing quality and easier operability with its non-contact and dry cutting mode, high-speed operation, zero TWR, and vibration. Because the thermal behavior of CFRP fibers and polymer matrix varies significantly, laser processing CFRP materials is distinct from other materials. Poor cut quality, a heat-affected zone (HAZ), and a taper angle can be attributed to a significant change in the decomposition temperature of the carbon fibers and matrix.

6.4.1. Mechanism to Material Removal Rate (MRR) Using Laser Machining

In laser beam cutting, a small laser light is used as the cutting tool to evaporate the material and move the erosion to the finished piece (Figure 6.5). As the vapors are expelled, the high-pressure (HP) assist gas moves the molten material, and the material is removed as the Kerf breadth increases. The anisotropic characteristics of FRPs, which are composed of two distinct components, complicate the LBM process mechanism for composite materials. Meanwhile, the physical and thermal characteristics of the matrix and the fiber reinforcement are very different. The low thermal conductivity matrix material will be removed by the chemical degradation process when the spot laser hits the object's surface during the LBM. Decomposition of the matrix happens when the chemical relations within it are melted or broken by the high-temperature laser beam. Since their thermal properties are quite comparable to the polymer matrix, this material removal method is particularly prevalent when machining thermoset polymer, wood, and aramid composites. Additionally, it takes longer and greater exposure time for the vaporization temperature to heat fibers with a high conductivity rapidly, such as carbon or glass fibers.

Figure 6.5. Mechanism of Laser machining process.

6.4.2. Evaluation of Laser Machining Efficiency (HAZ, Kerf Width, and Surface Roughness)

For thin laminates, where traditional drilling is inefficient, laser cutting stands out as a potential approach for machining FRP materials. Still, it is necessary to investigate and reduce several defects; the most critical of these is the width of the HAZ, which reduces the laminate mechanical strength. The two primary goals of laser machining are to keep the machining time to a minimum and to reduce heat damage to the cutting region. Factors such as gas pressure, frequency, laser power, cutting speed, and focus point position are the most important in laser machining. The most popular types of lasers are CO_2 and Nd: YAG. A CO_2 laser causes matrix recession and HAZ formation when it cuts FRP laminate because it introduces a high heat input into the material, which vaporizes a huge area of polymer resin. Because of their dissimilar physical, thermal, and mechanical characteristics, the amount of energy needed to break down fibers compared to resin material is different. A nonuniform heat absorption zone (i.e., HAZ) would also result from laser cutting GFRP because the orientation of the glass fibers causes heat conduction that is around ten times greater than that of the matrix. Laser cutting eliminates the resin surrounding the fibers, resulting in fiber extrusion because the workpiece temperature approaches the resin decomposition temperature while the fibers are still solid.

Table 6.3. Previous work on LBM of LPC/LPNC

Material	Thickness (mm)	Types of Laser & Technology	Process parameters	Response parameters	Techniques	Remarks	Ref.
Kelver, carbon fiber, and glass fiber	2mm for $C0_2$ and 0.5 mm for Excimer laser	Cutting and drilling with CO_2 and excimer laser	Focal length, focal spot	Surface quality and ablation rate	Finite differential model	The Kevlar matrix provides a dense and uniform structure, which is better than other composites. Excimer laser-processed rims are of higher grade.	(Dell'Erba et al., 1992)
Liquid crystalline polymer (LCP) composite	3 mm	Pulsed Nd: YAG and excimer Laser	Pulse energy, pulse duration, pulse frequency	Depth of hole	Mathematical model	The absorption of UV radiation by glass-fiber-reinforced LCP is greater than that of infrared light. A high-energy pulse or a repetitive number of pulses in YAG laser drilling generates carbonization and many surface cracks. In the excimer laser, such defects are also on the surface.	(Lau et al., 1994)
Three metals (aluminum, copper, mild steel) and CFRP	2 mm each and 1 mm for (CFRP)	Nd: YAG pulsed Laser	Pulse duration, Pulse Energy Beam radius, Peak beam energy flux	Hole shape, HAZ, and Fiber swelling	Numerical-based 3-D finite difference heat flow model	Numerical finite-based 2D and 3D models simulate the heat flow and phase change phenomena. Momentum transfer from the gas phase to the molten zone is primarily responsible for melt ejection. It is significantly less noticeable in composites.	(Cheng et al., 1998)
CFRP	2 mm	Pulsed Nd: YAG laser cutting	CS, PE, PD, (PRR), and GP	HAZ and Taper	RSM-based optimization	For most of the experimental factors in the middle range, the optimal parameter ranges were obtained. The main elements influencing cutting performance are the parent material's thermal properties and the fibers' volume fraction.	(Mathew et al., 1999)

Table 6.3. (Continued)

Material	Thickness (mm)	Types of Laser & Technology	Process parameters	Response parameters	Techniques	Remarks	Ref.
1080 type GFRP PCB material.	1.6 mm	UV YAG laser-drilled	Pulse repetition rate, laser power	HAZ	Parametric Analysis and Characterization	A clear hole wall with very little black charred material is generated when drilling with a lower laser power and repetition rate. Matrix recession and fiber protrusion are formed while drilling with a higher power and repetition rate. The effect of the pulse repetition rate for any given laser power is that a peak value of HAZ width is attained around 7 kHz, after which the HAZ width falls.	(Mathew et al., 1999)
Multi-layer carbon fiber composite	2 mm	Nd: YAG laser	Material thickness, pulse energy, and drilling time	Exit hole diameter and hole depth	Parametric Analysis	The diameter of the exit hole varies with pulse energy and material thickness. Drilling features provided for two distinct beam qualities, M^2=2.6 and 4.2.	(Rodden et al., 2002)
T300, High Modulus (HM) PAN (polyacrylonitrile) based fiber and P100 Pitch based fibers		Nd: YAG pulsed Laser	Laser power, Hole diameter, Pulse duration, Wave number	Measured hole diameter, Fiber Swelling, and Raman Intensity	2D finite difference numerical Heat flow model	SEM and Raman's spectroscopy were used to explore the influence of fiber type on the range of laser-induced fiber swelling. Compared to other fibers, extensive swelling and (HAZ) only occurred with low modulus poly acrylonitrile T300 (PAN)-based fibers. A 2D numerical heat flow model thermal field generation during laser drilling of carbon fibers is investigated.	(Voisey et al., 2006)

Material	Thickness (mm)	Types of Laser & Technology	Process parameters	Response parameters	Techniques	Remarks	Ref.
Thermoplastic sheet PMMA and Thermosets graphite fiber reinforced epoxy	2mm	Diode-pumped Nd: YAG laser	Wavelength, repetition frequency	Ra Heat affected zone Microcracks	Parametric Analysis	Only the UV beam can drill holes in thermoplastic PMMA and PC.	(Yalukova & Sárady, 2006)
Acrylic, PVC, and *PP* types of thermoplastics	2 mm	500W (CO_2) laser (Percussion drilling)	Number of gases, Standoff distance, Gas pressure	MRR, HT, Spatter thickness, and HC at exit and entry.	Taguchi's based orthogonal array (L_9)	It was feasible to control spatter production near the hole during LPD by adjusting the laser parameters. Acrylic, PVC, and PP are all thermoplastic polymers with nearly identical output performances.	(Masmiati & Philip, 2007)
CFRP	1.5 mm	Comparison between three laser sources: CO_2 lasers, Nd: YAG lasers, and Disk laser cutting	MP, PP, PF, CS, REPS, GP.	HAZ and Kerf width	Parameter variation	The cut quality of each laser system was optimized by adjusting the parameters under static tensile and bending situations. The quality was compared to traditional milling and abrasive water jet cut samples.	(Herzog et al., 2008)
UD- CFRP	0.3, 1, 3.1 and 7 mm	(DPSS) UV laser	Wavelength, pulse width, and laser power	HAZ	Mathematical modeling	Analysis of heat accumulation helps to avoid potential damage to CFRP. The BS shows similar results in laser drilling and mechanical drilling, while fiber cracking is found in the bottom of HAZ in laser drilling. The results reveal that the HAZ is kept to a minimum when cutting CFRP composites with a short-pulsed UV laser.	(X. Wang et al., 2018)

Table 6.3. (Continued)

Material	Thickness (mm)	Types of Laser & Technology	Process parameters	Response parameters	Techniques	Remarks	Ref.
CFRP	0.3 mm	355nm DPSS Nd: YVO_4 laser cutting	Scanning speed and scanning space	HAZ, ablation depth, Temperature distribution	3D Finite element modeling	FEA-based modeling illustrated CFRP composites' single- and double-line cutting employing heterogeneous meshing at different scanning speeds.	(Negarestani et al., 2010)
CFRP	1.2 mm	400W ns PDPSS Nd: YAG laser (MPS)	Temperature, Frequency, Pulse energy, Cutting speed.	Weight loss, MRR, Taper angle	RSM based on (CCD) design	A low volume fraction of oxygen at 12.5 percent combined with nitrogen at 8 bar assist gas pressure was the optimum process parameter for improving machining quality in laser cutting of CFRP. A mixture of oxygen and inert gas enhanced the vaporization or decomposition process to reduce heat damage.	(Monti & Nistico, 2008)
CFRP	2mm	Ps laser in PM	Wavelength, Average power, Pulse duration	HAZ	Parametric Analysis	A high-power Ps laser system with three different wavelengths (355, 532, and 1064 nm) was used for CFRP ablation. Examine the impact of various laser parameters on machining quality and processing time.	(Dubey & Yadava, 2008)
CFRP	2 mm	LPD with a *PS* laser system	Pulse energy, repetition rate	Extent of the HAZ	Parametric Analysis	The HAZ growth during laser processing was investigated with high temporal and spatial resolution. The MEZ was significantly reduced using high pulse energe and low repetition rates. Heat accumulation appears to be a significant effect when processing CFRP using pulsed laser systems.	(Onuseit et al., 2015)

Material	Thickness (mm)	Types of Laser & Technology	Process parameters	Response parameters	Techniques	Remarks	Ref.
CFRP	3 mm	Laser ablation, Shape-cutting technologies	LBP, PF, and wavelength	HAZ or Fd	Parametric analysis	Laser ablation of CFRP materials has much potential when using ultra-short laser pulses. Compared to shape-cutting technologies, laser processing is a modest manufacturing technique widely used for drilling and milling operations.	(Emmelmanna et al., 2011)
CFRP	3 mm	CO_2 laser cutting	Laser power, CS, PF, DC, GP	KW, TA, HAZ	Parametric Analysis	The effect of process parameters on cut quality was investigated in both CM and PM. A high-beam quality CO_2 laser in PM was used to cut with a minimum HAZ of roughly 540 μm.	(Dubey et al., 2012)
Aramid fiber, Glass fiber, Carbon fiber-based plastics	9.5 mm	High-power Ps/Fs-based lasers and Diode pumped solid state lasers	pulse length, wavelength, PD, SS, and PG.	HAZ, kerf widths, and depth of cut.	Parametric Analysis	Laser machining of composite materials requires short pulses and high scanning speed while requiring short wavelengths to minimize heat effects. Laser machining is slower than mechanical or water jet cutting methods, but it avoids some issues, including TWR and water penetration.	(Soliman et al., 2012)
Carbon black/vinylester/ glass laminates	3 mm	300 W CO_2 Laser	Laser power, pulse frequency, Scanning speed, and carbon black (wt.%)	HAZ and TA	Taguchi-based L_{16} orthogonal array.	HAZ and TA were reduced by adding 6% carbon black into vinyl-ester/glass laminates, while HAZ was lowered using a lower laser power. The Taper angle was reduced by using a greater laser power. For multiple responses, GRA was used to determine the best combination of laser drilling parameters.	(Nagesh et al., 2013)

Table 6.3. (Continued)

Material	Thickness (mm)	Types of Laser & Technology	Process parameters	Response parameters	Techniques	Remarks	Ref.
CFRP	1.2 mm	1 kW fiber laser cutting	LP, SS, GP, FPP	Kerf width, matrix regression at beam inference	RSM-based CCD design	The MPC technique effectively minimizes Fd at low power and fast scanning rates compared to SPC technology.	(Leone et al., 2013)
CFRP	0.5 mm thick	Yb: YAG fiber laser cutting, MPST	Pulse Power, scanning speed	CS, kerf geometry, and HAZs	Parametric Analysis	MPST is useful for obtaining a small kerf and limiting HAZs. The scanning speed and pulse power determine the effective cutting speed, whereas the scanning speed mostly determines the kerf geometry. The HAZ's extent is affected by scanning speed and pulse power.	(Bluemel et al., 2014)
CFRP and GFRP	3 mm	CO_2 Laser cutting	LBP, CS, GP, Workpiece material	MRR, Ra, kerf width	Parametric Analysis	On the CFRP and GFRP samples, charring and melting of the matrix phase and surface voids and cavities were observed. HAZ was apparent in most CFRP specimens, although minimal fiber protrusion was visible in GFRP.	(Shyha et al., 2013)
CFRP material	3 mm	Q-switch Nd: YAG laser, laser-drilled holes for 0.4, 0.6, 0.8 mm Diameter	Frequency, thickness, and hole diameter	(HAZ)	Mathematical modeling	A short-pulsed laser was used to reduce HAZ during CFRP micromachining. SEM analysis clearly shows proper layer bonding in an excellent completed micro hole.	(Gräf et al., 2015)

Material	Thickness (mm)	Types of Laser & Technology	Process parameters	Response parameters	Techniques	Remarks	Ref.
Carbon black/vinylester/ glass laminates	3mm	300 W CO_2 Laser	Cutting speed, Laser power, air pressure, and addition of nanofiller content.	Damage width, HAZ and TA	L_9 Orthogonal Array using Grey Relational Analysis.	The effect of nanofiller addition on the quality of laser drilling of vinyl ester/glass was examined. Three nanofiller materials were studied: nickel nanopowder, carbon black, and Closite 15-A. Because of the absence of char, nickel nanopowder substantially reduced HAZ and enhanced hole quality. Second-order regression models were developed to predict the optimal value of laser drilling responses.	(Staupendahl, 2014)
CFRP	7 layers with $s = 2.2$ mm thickness	Cutting, ablation, and drilling by 70W SP nanosecond laser, Pulsed disk lasers	Frequency	HAZ, Kerf width	Parametric Analysis	GST techniques produced excellent surface and edge qualities in terms of Fd-free interface when drilling and cutting processes were applied in multi-cycle mode. Nanosecond disk laser machining was a suitable tool for cutting and drilling operations in the automotive and aircraft industries.	(Lin et al., 2015; Niino et al., 2015)
(CFRTP) and CFRP	3 mm and 2 mm	2D/3D laser Cutting by cw IR fiber laser (MCPIT) Technique	Number of laser passes	Cutting depth	Parametric Analysis	Laser irradiation in an MCPI Technique removed debris from the grooves of CFRTP and CFRP. Laser cutting with high-speed beam scanning produces a clean top and good sidewall quality while generating the least amount of HAZ. A five-axis laser cutting machine is used to cut a 3-D CFRP plate.	(Niino et al., 2015)

Table 6.3. (Continued)

Material	Thickness (mm)	Types of Laser & Technology	Process parameters	Response parameters	Techniques	Remarks	Ref.
CFRP	1, 2, 6 mm, three different thickness	Laser milling and drilling using 400W picosecond laser. Single-ring trepanning strategy.	Scanning speed, laser power, Ring spacing, and repetition time	HAZ and ablation depth	OPAT	When picosecond was applied in place of a nanosecond, the taper angle and HAZ were reduced to 25µm. The holes generated in 6-mm-thick CFRP had a maximum taper angle of roughly 15^0 degrees. As the laser power and scanning speed increase, HAZ and ablation depths decrease.	(Salama, Li, et al., 2016)
CFRP and mild steel	6 mm (CFRP) and 2 mm (mild steel), Self-limiting, Machining	Using picosecond Single and multiple ring trepanning strategy	Number of rings, machining time, peak power, and wavelength	Taper angle, kerf width, ablation depth, and MRR	Mathematical simulation	The multiple-ring technique led to a greater ablation depth and MRR than single-ring trepanning.	(Salama, Yan, et al., 2016)
MWCNT-doped epoxy/GF polymeric nanocomposite	5.40 mm.	Continuous mode (CW) fiber laser	MWCNT (wt%), Laser Power, Scan speed, energy per unit length	HAZ, TA, and SR	Parametric Analysis	SAM is a novel technique for examining laser drilling damage near the hole by offering ply-by-ply damage. EDX analysis provided the elemental analysis of the burned debris deposited on the hole wall. The addition of multiwalled carbon nanotubes to the polymer matrix significantly improved the hole quality.	(Kumar & Singh, 2016)

Material	Thickness (mm)	Types of Laser & Technology	Process parameters	Response parameters	Techniques	Remarks	Ref.
GFRP	5 mm	Fiber laser cutting	LP, CS, and GP	HAZ	ANN-based regression model.	Using regression and ANN-based modeling, HAZ grows with increasing laser power and cutting speed. When simulated and experimental data are compared, the ANN model performs better in predicting HAZ for the given range of input parameters.	(Patel et al., 2016)
CFRP	1.4 mm	Fiber Laser Cutting	Laser Power, Cutting Speed, Gas Flow Rate	Kerf Width, HAZ, Taper in %	RSM-based CCD modeling and optimization	The optimum values are laser power 260 watts, CS 4500 mm per min, and gas flow rate 14.23 l/min. The responses are kerf width, taper percentage, and the width of HAZ, which were 163.7 µm, 5.75%, and 573.28 µm, respectively. The confirmation experiments were conducted, and results were provided at 2.70%, 1.87%, and 0.36% for kerf width, taper percentage, and width of HAZ, respectively.	(Herzog et al., 2016)
Glass/epoxy composite laminated with copper used in PCB.	Glass/epoxy composite is 25 μm. And copper coating 2 µm	Picosecond IR pulsed laser	Laser powers, Repetition rates, and elapsed times	Top- and bottom-hole diameter, Smear zone, and taper ratio	Parametric Analysis and Characterization	Ps-IR based pulsed Laser generated sub-50-µm dia as BVHs on copper-coated glass/epoxy. Laser drilling has been modified to generate a Blind via hole (BVH) to connect PCB layers. The total fluence is calculated by multiplying pulse energy and number of shots and dividing it by the beam spot size. $F_{total}= (nE_{pulse}/d)$	(Samsingh et al., 2017)

Table 6.3. (Continued)

Material	Thickness (mm)	Types of Laser & Technology	Process parameters	Response parameters	Techniques	Remarks	Ref.
CFRP is made from PAN-based carbon fiber	4.5 mm	CW mode laser of wavelength 1030 nm	Avg. Laser power, drilling time	Pressure ratio and flow speed	Parametric Analysis	High-speed imaging was used to examine the hot plume of ablation products produced during the laser drilling of CFRP with a CW laser beam. With increased average power, the flow speed increases. Compression shocks can also be seen when a moving laser beam is used to cut CFRP.	(Faas et al., 2017)
CFRP 3D Cs display	2mm	Nd: YAG laser beam	energy density, pulse energy, mean power	damage analysis in terms of the crater.	Mathematical Analysis.	ImageJ software was utilized to analyze damages based on SEM and LMM quantitatively. Crater-like damages were produced on the surface of the composite due to absorbed thermal energy.	(Lewicki et al., 2017)
CFRP	2.2mm	Laser cutting, Laser drilling by Ns pulsed high-power disk laser	Laser repetition rates, Scanning speed, Delay time, number of passes, Gas pressure	HAZ	Parametric Analysis	During the cooling technique, supplying gas flow at the specimen's top and bottom surfaces helps lower the thermal load during CFRP laser processing. The effectiveness was determined by simultaneously monitoring the temperatures on the top and bottom surfaces of the material, as well as measuring the HAZ on cross-section micrographs.	(Rehbein et al., 2017)

Material	Thickness (mm)	Types of Laser & Technology	Process parameters	Response parameters	Techniques	Remarks	Ref.
CFRP pieces	5.85 mm	Fiber laser doped with Ytterbium (Yb)	Pulsed frequency and Scanning speed,	Hole taper, Ra, HAZ	Parametric Analysis	ANOVA showed the influence of scanning speed and frequency on surface quality. The techniques of SOM and SEM were utilized to find defects in the drill wall.	(Mayuet Ares et al., 2018)
GFRP	1.2 mm	CO_2 laser in CW mode, Single pass trepanning drilling method	Gas pressure, cutting speed, and laser intensity	HAZ, TA, Ra and tensile strength for conventional drilling	Full factorial design generated by Minitab 17 software	Low cutting speed and high laser intensity negatively impacted SR, HAZ, and TS, while high laser intensity helped minimize the taper angle. The optimization helped reduce SR from 5.35μm in traditional drilling to 3.1μm in laser drilling.	(Schneider & Petring, 2018)
MWCNT-doped epoxy/CF polymeric nanocomposite	4.0 mm	(CW) Fiber laser	MWCNT (wt%), Power, Scan speed, and Energy per unit length	HAZ, TA, and SR	Parametric Analysis	The thermal diffusivity of the polymer matrix is increased due to the high thermal conductivity of multiwalled carbon nanotubes. A new technique, scanning acoustic microscopy, was used to investigate the role of multiwalled carbon nanotubes on the outcome of thermal damage. The results showed that 1.5 wt percent multiwalled CNT-doped carbon fiber-reinforced polymer laminates reduced the HAZ, TA, and SR of holes by 30%, 47%, and 43%, respectively.	(W. Li et al., 2018)

Table 6.3. (Continued)

Material	Thickness (mm)	Types of Laser & Technology	Process parameters	Response parameters	Techniques	Remarks	Ref.
GFRP based (PA6/GF60) and a CFRP based (PA6/CF60) composite	1.7 to 2.8 mm	Single-mode Fiber laser cutting	Maximum laser power, Wavelength, Fiber diameter, Maximum travel, Positioning accuracy, Repeatability, and Maximum speed.	HAZ, Mechanical properties analysis	Parametric Analysis.	TGA exposed decomposition temperatures of 457.6°C for the PA6 matrix and 737.3°C for carbon fibers, while no thermal degradation of the glass fibers was identified up to 1,000^0 C. Compared to PA6/CF60 composites, single-mode fiber laser cutting required more cutting cycles. The mechanical properties of laser-cut composites are equivalent to water jet-cut composites.	(Benmokrane et al., 2018)
GFRP	4.56 mm	CW mode Fiber laser with 400W power	LI, CS, GP	Kerf width, HAZ, Taper in percentage	RSM-based CCD modeling and optimization	The optimum laser irradiance and cutting speed values were 8.28 watts/cm^2×10^5 and 600 mm/min at an assist gas pressure of 7.84 bar. The responses were upper kerf width: 177.4 mm, taper 0.73%, and HAZ on the top surface: 109.23 mm, respectively. The confirmation experiments were conducted, and the results were 2.52%, 1.84%, and 0.45 % for upper kerf width, taper percentage, and top surface HAZ, respectively.	(B et al., 2019)

Material	Thickness (mm)	Types of Laser & Technology	Process parameters	Response parameters	Techniques	Remarks	Ref.
Basalt–glass hybrid composite	2.23 mm	Pulsed Nd: YAG laser	Air pressure, pulse frequency, and lamp current	HC and HAZ	RSM-based BBD design.Mathematical Modelling using ANN-based (MOGA)	The percentage discrepancies between predicted and experimental values of HC and HAZ are determined to be 2.291 percent and 7.23 percent, respectively, utilizing ANN-based optimization techniques MOGA.	(Chen et al., 2019)
GFRP	2.5 mm	Continuous mode CO2 laser, Single pass trepanning drilling method	Gas pressure, cutting speed and laser power, Feed rate, spindle speed For conventional drilling	HAZ, bearing test	A full factorial design based on GA-ANN.	By increasing the HAZ, the bearing strength was reduced nearly linearly. Laser drilling achieves higher bearing strength than mechanical drilling. The GA-ANN-based optimization enhanced the predictive model performance R2 by 5–34 percent compared to the trial-and-error method.	(Solati et al., 2019)
Epoxy-glass printed wiring boards (PWBs)	0.5 mm PWB clad with 18μm copper coil	Short-pulse CO_2 laser	Laser pulse duration and frequency	Recast layer thickness and heat-affected zone	Parametric Analysis, Generate heat conduction model	CO2 laser drilling technology can create micro vias in epoxy-glass printed wiring boards (PWBs). When the pulse frequency increased, the carbonized layer thickness around the drilled holes increased significantly to 40 to 60 Hz. According to the simple heat conduction model, the thickness of the heat-affected layer was reduced from 8 μm to 2.6 μm by reducing the pulse duration from 500 μs to 50μs.	(Sohaili et al. 2019)

Table 6.3. (Continued)

Material	Thickness (mm)	Types of Laser & Technology	Process parameters	Response parameters	Techniques	Remarks	Ref.
CFRP	1.8 mm	High-power fiber laser in pulse mode	Wavelength, Average power, Modulation frequency,	HAZ and Fd factor	Parametric analysis	The DIC approach revealed that crack propagation and failure mode are related to the high level of stresses induced around the hole. HAZ at the hole's entry and exit was reduced as cutting speed increased, although the duty cycle had a minimum impact on HAZ.	(Heiderscheit et al., 2019)
GFRP		CO_2 laser engraving	P/U, depth along the fiber and across the fiber	SR along the fiber direction and across the fiber direction.	Parametric Analysis, SEM imaging	Experiments were conducted at five distinct ED settings and three different workpiece defocus points. A semi-empirical depth estimation model was developed along and across the fiber engraving. CO_2 laser engraving of GFRP is an effective method for small-scale material removal. The engraving depth along the fiber was greater than across at the same ED.	(Prakash, 2019)
CFRP	1 mm	Fiber optic-delivered 1000 W laser power	Peak Power, Peak Power Density, Pulse Frequency, and duty cycle	Mean penetration time, entrance hole diameter, surface HAZ width, material ablation depth, hole taper	Numerical simulation	This experiment used a near-infrared (NIR) laser drilling of a CFRP composite in CW mode and a ms pulse mode laser. The laser penetration time dropped sharply when the peak power density exceeded 23.4 kW/cm2.	(X. Li et al., 2020)

Material	Thickness (mm)	Types of Laser & Technology	Process parameters	Response parameters	Techniques	Remarks	Ref.
CFRP	25.4 mm	IPG 1kW Fiber Laser by spiral trepanning strategy	Laser Power, Scanning Speed, Pulse-on Time, and Pulse repetition rate	Hole depth, HAZ	Parametric Analysis	Fiber is a unique laser drilling technique that penetrates a thick CFRP to a depth of 22 mm.	(Ahmad Sobri et al., 2020)
CFRP	10 mm	Coaxial trepan drilling New (DBOD) Techniques	Laser pulse heat flux, depth of drilling, defocus depth	Material removal mechanisms and (HAZ)	3-D heat flow model generated by FEM	A comparative analysis between the two techniques, DBOD and single-beam process, is needed for a thick CFRP plate. The DBOD drilling process improves processing efficiency by two times.	(Tao et al., 2020)
Aramid/epoxy and glass/ epoxy in PWB	The core thickness is 0.8 mm, and the build-up layer is 0.3 mm thick.	CO_2 laser with 100 W power	Irradiation time, interval time, number of pulses	Damage width, drill depth, taper angle	Parametric Analysis	Maintaining a 1-second interval between irradiations can avoid heat accumulation when drilling with numerous irradiations. When Aramid fiber is used, the hole-drilling efficiency is twice that of glass cloth. The taper angle of the hole wall improves plating in the construction of a circuit while Laser drilling a blind hole.	(Oguz et al., 2021)

In addition, the heat produced is substantially higher than the resin matrix vaporization temperature when the temperature approaches the glass fiber vaporization point. So, in the HAZ, which is located close to the cutting kerf, the resin is either melted or evaporated. In addition, the machined surface exhibits a lower HAZ perpendicular to the fiber orientation and a greater HAZ along the fiber orientation due to the discrepancy in the heat conduction coefficients of the matrix and the fibers. Various methods have been investigated to enhance the laser cut surface characteristics, including HAZ, Ra, taper angle, and kerf width when cutting laminates, as summarized in the report in Table 6.3.

The table clearly shows that the most critical factors when laser drilling CFRP composites are HAZ, taper angle, and machined surface quality. As usual, the disparity between the carbon fiber and epoxy resin decomposition temperatures has led numerous researchers to the conclusion that thermal damage is the more pressing issue. In addition, following laser drilling, GFRPs showed less machining damage compared to CFRPs. Also, some recent research has looked at how improving the surface properties of laser-drilled/machined FRP laminates by introducing various nanomaterials affects those laminates. According to the study, the HAZ significantly impacts the drilled FRP laminates' bearing strength and hole quality. Continuous research is required to explore and improve the mechanical attributes of FRP laminates and surface quality standards. Composites have replaced metals for numerous uses where weight is an issue, such as in the aerospace, defense, and automobile industries. A polymer matrix and an arrangement of fibers are typically combined to form a composite material. Composites degrade and fail during conventional machining due to reinforcing fiber Fd from the matrix, which compromises the material characteristics. Because of the high fiber strength, conventional machining of composites uses expensive machine equipment at a rapid rate of TWR. It can achieve HAZ-free output for atypical machining choices, such as water jet and AWJM. However, note that the composite will get saturated and eventually collapse when subjected to strain. Abrasive water jet cutting is not a good fit for the role because of the abrasive particle deposition and slow machining speed. On the other hand, laser cutting eliminates the need for complicated fixtures, uses a focused micron-size laser beam to quickly machine intricate shapes and precise geometries with low kerfs, and does not entail mechanical cutting force or TWR. Due to the notable disparity in the thermal characteristics between the matrix and the fiber, they each exhibit distinct reactions to the laser source. Therefore, the selection of laser cutting parameters, including laser pulse width, laser power, cutting speed, and laser pulse frequency, is crucial in the laser machining of composites due to the material anisotropy. Precise micro-manufacturing applications benefit from pulsed lasers because their intensity modulation reduces heat damage to the surrounding material. Due to energy dissipation differences, the matrix and fiber remove material improperly despite having similar absorbance at the CO2 laser wavelength. Heat diffusivity and vaporization time differences result in large heat-affected zones and Fd officers. This can be solved with pulse lasers and shorter wavelengths. A 1 mm thick FRP plate was cut at a speed of 500 mm/s with a Kerf width of around 0.2 mm using a pulsed Nd: YAG laser source. The laser had a fundamental wavelength of 1064 nm and an average power of 100 W. Moreover, with reduced HAZ, the cut quality of the FRP composites was enhanced.

Researchers looked at how the cutting speed affects cut quality and found that the heat-affected zone diminished as cutting speed increased. Because the laser beam had less time to interact with the material at faster cutting speeds, the kerf width at the entrance and exit decreased. With a compromise in machining time, picosecond and femtosecond lasers decrease the HAZ to 10 - 25 μm. High repetition ultra-short pulsed lasers might improve the machining time, but the HAZ would increase with an increased laser spotlight. Considering the foregoing, further research is required into the intricate relationship between laser and material properties and HAZ. There has been a recent increase in research on hybrid materials, which enhance strength and impact resistance by mixing reinforcements with inorganic particles incorporated into the combined polymer matrix. Polymer nanocomposites differ from filled plastics and conventional composites due to their large internal interfacial area and anisotropic dimensions between components. Machining polymer nanocomposites, however, have received fewer studies.

Conclusion

AWJM, LBM, and EDM are three of the most prominent non-traditional machining technologies that are now utilized to cut LPC. On the other hand, each machining technique has some constraints when it comes to machine FRPs. No HAZ forms on the machined composite during AWJM because it is a cold machining technique. However, the hybrid material's usual performance has been drastically diminished due to extensive laminar Fd, particularly on the bottom kerf. LBM offers versatility and the capacity to cut hard-to-machine materials at fast speeds as an alternative to machining composites. LBM produces the most machinability output. However, hot cutting causes HAZ and severe surface marks. As an alternative, the EDM technique was used because of its accuracy and capacity to provide a machined surface devoid of burrs without the need for cutting force.

Nevertheless, among the machining techniques, EDM offers comparatively poor MMR and is only feasible for materials with high electrical conductivity, like carbon fiber. Parameter settings for AWJM, LBM, and EDM procedures can be carefully chosen to minimize machining damage. On the other hand, another crucial topic not explored in previous research is the multi-response optimization of these machining processes.

References

Ablyaz, T. R., Shlykov, E. S., & Muratov, K. R. (2021). Improving the Efficiency of Electrical Discharge Machining of Special-Purpose Products with Composite Electrode Tools. *Materials*, *14*(20). https://doi.org/10.3390/ma14206105.

Ahmad Sobri, S., Heinemann, R., & Whitehead, D. (2020). Development of Laser Drilling Strategy for Thick Carbon Fibre Reinforced Polymer Composites (CFRP). *Polymers*, *12*(11). https://doi.org/10.3390/polym12112674.

Anu Kuttan, A., Rajesh, R., & Dev Anand, M. (2021). Abrasive water jet machining techniques and parameters: a state of the art, open issue challenges and research directions. In *Journal of the Brazilian Society of Mechanical Sciences and Engineering* (Vol. 43, Issue 4). Springer Science and Business Media Deutschland GmbH. https://doi.org/10.1007/s40430-021-02898-6.

B, S. D. R., Sethi, A., & Das, A. K. (2019). Fiber laser processing of GFRP composites and multi-objective optimization of the process using response surface methodology. *Journal of Composite Materials*, *53*(11), 1459–1473. https://doi.org/10.1177/0021998318805139.

Benmokrane, B., Nazair, C., Loranger, M.-A., & Manalo, A. (2018). Field Durability Study of Vinyl-Ester–Based GFRP Rebars in Concrete Bridge Barriers. *Journal of Bridge Engineering*, *23*(12), 4018094. https://doi.org/10.1061/(ASCE)BE.1943-5592.0001315.

Bluemel, S., Jaeschke, P., Suttmann, O., & Overmeyer, L. (2014). Comparative Study of Achievable Quality Cutting Carbon Fibre Reinforced Thermoplastics Using Continuous Wave and Pulsed Laser Sources. *Physics Procedia*, *56*, 1143–1152. https://doi.org/https://doi.org/10.1016/j.phpro.2014.08.029.

Chen, D., Luo, Q., Meng, M., Li, Q., & Sun, G. (2019). Low velocity impact behavior of interlayer hybrid composite laminates with carbon/glass/basalt fibres. *Composites Part B: Engineering*, *176*, 107191. https://doi.org/https://doi.org/10.1016/j.compositesb.2019.107191.

Cheng, C. F., Tsui, Y. C., & Clyne, T. W. (1998). Application of a three-dimensional heat flow model to treat laser drilling of carbon fibre composites. *Acta Materialia*, *46*(12), 4273–4285. https://doi.org/https://doi.org/10.1016/S1359-6454(98)00090-1.

Deepa, A., Kuppan, P., & Krishnan, P. (2020). *A Comparative Study of Structural Changes in Conventional and Unconventional Machining and Mechanical Properties Evaluation of Polypropylene Based Self Reinforced Composites*. *27*(1), 108–118. https://doi.org/doi:10.1515/secm-2020-0012.

Dell'Erba, M., Galantucci, L. M., & Miglietta, S. (1992). An experimental study on laser drilling and cutting of composite materials for the aerospace industry using excimer and CO2 sources. *Composites Manufacturing*, *3*(1), 14–19. https://doi.org/https://doi.org/10.1016/0956-7143(92)90178-W.

Dubey, A. K., Norkey, G., & Agrawal, S. (2012). Parameter optimisation in laser cutting of aluminium alloy sheet. In *Int. J. Mechatronics and Manufacturing Systems* (Vol. 5).

Dubey, A. K., & Yadava, V. (2008). Laser beam machining—A review. *International Journal of Machine Tools and Manufacture*, *48*(6), 609–628. https://doi.org/https://doi.org/10.1016/j.ijmachtools.2007.10.017.

Dutta, Hrishikesh, Debnath, Kishore, & Sarma, Deba Kumar. (2021). Investigation on cutting of thin carbon fiber-reinforced polymer composite plate using sandwich electrode-assisted wire electrical-discharge machining. *Proceedings of the Institution of Mechanical Engineers, Part E*, *235*(5), 1628–1638. https://doi.org/10.1177/09544089211013318.

Emmelmanna, C., Petersen, M., Goeke, A., & Canisius, M. (2011). Analysis of Laser Ablation of CFRP by Ultra-Short Laser Pulses with Short Wavelength. *Physics Procedia*, *12*, 565–571. https://doi.org/https://doi.org/10.1016/j.phpro.2011.03.071.

Faas, S., Freitag, C., Boley, S., Berger, P., Weber, R., & Graf, T. (2017). Flow speed of the ablation vapors generated during laser drilling of CFRP with a continuous-wave laser beam. *Applied Physics A*, *123*(3), 156. https://doi.org/10.1007/s00339 017-0781-0.

Gopal, P. M., Kavimani, V., & Arunkumar, K. (2023). Multi-objective optimization on abrasive water jet machining of epoxy/glass fiber/ grinding wheel particle composite through hybrid optimization technique. *Multiscale and Multidisciplinary Modeling, Experiments and Design*, *6*(4), 697–707. https://doi.org/10.1007/s41939-023-00177-6.

Gräf, S., Staupendahl, G., Krämer, A., & Müller, F. A. (2015). High precision materials processing using a novel Q-switched CO2 laser. *Optics and Lasers in Engineering*, *66*, 152–157. https://doi.org/https://doi.org/10.1016/j.optlaseng.2014.09.007.

Heiderscheit, T., Shen, N., Wang, Q., Samanta, A., Wu, B., & Ding, H. (2019). Keyhole cutting of carbon fiber reinforced polymer using a long-duration nanosecond pulse laser. *Optics and Lasers in Engineering*, *120*, 101–109. https://doi.org/https://doi.org/10.1016/j.optlaseng.2019.03.009.

Herzog, D., Jaeschke, P., Meier, O., & Haferkamp, H. (2008). Investigations on the thermal effect caused by laser cutting with respect to static strength of CFRP. *International Journal of Machine Tools and Manufacture*, *48*(12), 1464–1473. https://doi.org/https://doi.org/10.1016/j.ijmachtools.2008.04.007.

Herzog, D., Schmidt-Lehr, M., Oberlander, M., Canisius, M., Radek, M., & Emmelmann, C. (2016). Laser cutting of carbon fibre reinforced plastics of high thickness. *Materials & Design*, *92*, 742–749. https://doi.org/https://doi.org/10.1016/j.matdes.2015.12.056.

Hutyrová, Z., Ščučka, J., Hloch, S., Hlaváček, P., & Zeleňák, M. (2016). Turning of wood plastic composites by water jet and abrasive water jet. *The International Journal of Advanced Manufacturing Technology*, *84*(5), 1615–1623. https://doi.org/10.1007/s00170-015-7831-6.

Jagadish, Bhowmik, S., & Ray, A. (2016). Prediction and optimization of process parameters of green composites in AWJM process using response surface methodology. *International Journal of Advanced Manufacturing Technology*, *87*(5–8), 1359–1370. https://doi.org/10.1007/s00170-015-8281-x.

Jagadish, Gupta, K., & Rajakumaran, M. (2018). Evaluation of machining performance of pineapple filler based reinforced polymer composites using abrasive water jet machining process. *IOP Conference Series: Materials Science and Engineering*, *430*(1). https://doi.org/10.1088/1757-899X/430/1/012046.

Jain, V. K. (2008). Advanced (Non-traditional) Machining Processes. In J. P. Davim (Ed.), *Machining: Fundamentals and Recent Advances* (pp. 299–327). Springer London. https://doi.org/10.1007/978-1-84800-213-5_11.

Jani, S. P., Kumar, A. S., Khan, M. A., & Kumar, M. U. (2016). Machinablity of Hybrid Natural Fiber Composite with and without Filler as Reinforcement. *Materials and Manufacturing Processes*, *31*(10), 1393–1399. https://doi.org/10.1080/10426914.2015.1117633.

Kalirasu S, Rajini N, Rajesh S, Winowlin Jappes J. T, & Karuppasamy K. (2016). AWJM Performance of Jute/Polyester Composite using MOORA and Analytical Models. *Materials and Manufacturing Processes*, *32*(15), 1730–1739.

Kalusuraman, G., Thirumalai Kumaran, S., Siva, I., & Anand Kumar, S. (2020). Cutting Performance of Luffa Cylindrica Fiber–Reinforced Composite by Abrasive Water Jet. *Journal of Testing and Evaluation*, *48*(5), 3417–3428. https://doi.org/10.1520/JTE 20180330.

Kechagias, J., Petropoulos, G., & Vaxevanidis, N. (2012). Application of Taguchi design for quality characterization of abrasive water jet machining of TRIP sheet steels. *International Journal of Advanced Manufacturing Technology*, *62*(5–8), 635–643. https://doi.org/10.1007/s00170-011-3815-3.

Kumar, D., & Singh, K. K. (2016). An experimental investigation of surface roughness in the drilling of MWCNT doped carbon/epoxy polymeric composite material. *IOP Conference Series: Materials Science and Engineering*, *149*(1), 12096. https://doi.org/10.1088/1757-899X/149/1/012096.

Lau, W. S., Wang, M., & Lee, W. B. (1990). Electrical discharge machining of carbon fibre composite materials. *International Journal of Machine Tools and Manufacture*, *30*(2), 297–308. https://doi.org/https://doi.org/10.1016/0890-6955(90)90138-9.

Lau, W. S., Yue, T. M., & Jiang, C. Y. (1994). Machining behaviours of a liquid crystal polymer composite using Nd:YAG and Excimer lasers. In E. Usui (Ed.), *Advancement of Intelligent Production* (pp. 570–575). Elsevier. https://doi.org/https://doi.org/10.1016/B978-0-444-81901-7.50109-2.

Leone, C., Papa, I., Tagliaferri, F., & Lopresto, V. (2013). Investigation of CFRP laser milling using a 30W Q-switched Yb:YAG fiber laser: Effect of process parameters on removal mechanisms and HAZ formation. *Composites Part A: Applied Science and Manufacturing*, *55*, 129–142. https://doi.org/https://doi.org/10.1016/j.compositesa.2013.08.004.

Lewicki, J. P., Rodriguez, J. N., Zhu, C., Worsley, M. A., Wu, A. S., Kanarska, Y., Horn, J. D., Duoss, E. B., Ortega, J. M., Elmer, W., Hensleigh, R., Fellini, R. A., & King, M. J. (2017). 3D-Printing of Meso-structurally Ordered Carbon Fiber/Polymer Composites with Unprecedented Orthotropic Physical Properties. *Scientific Reports*, *7*(1), 43401. https://doi.org/10.1038/srep43401.

Li, W., Xiang, D., Wang, L., Harkin-Jones, E., Zhao, C., Wang, B., & Li, Y. (2018). Simultaneous enhancement of electrical conductivity and interlaminar fracture toughness of carbon fiber/epoxy composites using plasma-treated conductive thermoplastic film interleaves. *RSC Adv.*, *8*(47), 26910–26921. https://doi.org/10.1039/C8RA05366A.

Li, X., Hou, W., Han, B., Xu, L., Li, Z., Nan, P., & Ni, X. (2020). Investigation on the Continuous Wave Mode and the ms Pulse Mode Fiber Laser Drilling Mechanisms of the Carbon Fiber Reinforced Composite. *Polymers*, *12*(3). https://doi.org/10.3390/polym12030706.

Lin, Y., Gigliotti, M., Lafarie-Frenot, M. C., Bai, J., Marchand, D., & Mellier, D. (2015). Experimental study to assess the effect of carbon nanotube addition on the through-thickness electrical conductivity of CFRP laminates for aircraft applications. *Composites Part B: Engineering*, *76*, 31–37. https://doi.org/https://doi.org/10.1016/j.compositesb.2015.02.015.

Masmiati, N., & Philip, P. K. (2007). Investigations on laser percussion drilling of some thermoplastic polymers. *Journal of Materials Processing Technology*, *185*(1–3), 198–203. https://doi.org/10.1016/j.jmatprotec.2006.03.151.

Mathew, J., Goswami, G. L., Ramakrishnan, N., & Naik, N. K. (1999). Parametric studies on pulsed Nd:YAG laser cutting of carbon fibre reinforced plastic composites. *Journal of Materials Processing Technology*, *89–90*, 198–203. https://doi.org/https://doi.org/10.1016/S0924-0136(99)00011-4.

Mayuet Ares, P. F., Vázquez Martínez, J. M., Marcos Bárcena, M., & Gámez, A. J. (2018). Experimental Study of Macro and Microgeometric Defects in Drilled Carbon Fiber Reinforced Plastics by Laser Beam Machining. *Materials*, *11*(8). https://doi.org/10.3390/ma11081466.

Mesalamy, A. S. E., & Youssef, A. (2020). Enhancement of cutting quality of abrasive waterjet by using multipass cutting strategy. *Journal of Manufacturing Processes*, *60*, 530–543. https://doi.org/10.1016/j.jmapro.2020.10.036.

Monti, G., & Nistico, N. (2008). Square and rectangular concrete columns confined by CFRP: Experimental and numerical investigation. *Mechanics of Composite Materials*, *44*(3), 289–308. https://doi.org/10.1007/s11029-008-9021-1.

Nagesh, S., Narasimha Murthy, H. N., Krishna, M., & Basavaraj, H. (2013). Parametric study of CO2 laser drilling of carbon nanopowder/vinylester/glass nanocomposites using design of experiments and grey relational analysis. *Optics & Laser Technology*, *48*, 480–488. https://doi.org/https://doi.org/10.1016/j.optlastec.2012.11.013.

Natarajan, Y., Murugesan, P. K., Mohan, M., & Liyakath Ali Khan, S. A. (2020). Abrasive Water Jet Machining process: A state of art of review. In *Journal of Manufacturing Processes* (Vol. 49, pp. 271–322). Elsevier Ltd. https://doi.org/10.1016/j.jmapro.2019.11.030.

Negarestani, R, Sundar, M, Sheikh, M A, Mativenga, P, Li, L, Li, Z L, Chu, P L, Khin, C C, Zheng, H Y, & Lim, G C. (2010). Numerical simulation of laser machining of carbon-fibre-reinforced composites. *Proceedings of the Institution of Mechanical Engineers, Part B, 224*(7), 1017–1027. https://doi.org/10.1243/09544054JEM1662.

Niino, H., Harada, Y., Anzai, K., Aoyama, M., Matsushita, M., Furukawa, K., Nishino, M., Fujisaki, A., Miyato, T., & Kayahara, T. (2015). 2D/3D laser cutting of carbon fiber reinforced plastic (CFRP) by fiber laser irradiation. *Proc.SPIE, 9353*, 935303. https://doi.org/10.1117/12.2077447.

Oguz, Z. A., Erklig, A., & Bozkurt, Ö. Y. (2021). Degradation of hybrid aramid/glass/epoxy composites hydrothermally aged in distilled water. *Journal of Composite Materials, 55*(15), 2043–2060. https://doi.org/10.1177/0021998320984237.

Onuseit, V., Freitag, C., Wiedenmann, M., Weber, R., Negel, J.-P., Löscher, A., Ahmed, M. A., & Graf, T. (2015). Efficient processing of CFRP with a picosecond laser with up to 1.4 kW average power. *Proc. SPIE, 9350*, 935012. https://doi.org/10.1117/12.2077457.

Patel, P., Sheth, S., & Patel, T. (2016). Experimental Analysis and ANN Modelling of HAZ in Laser Cutting of Glass Fibre Reinforced Plastic Composites. *Procedia Technology, 23*, 406–413. https://doi.org/https://doi.org/10.1016/j.protcy.2016.03.044.

Popan, I. A., Cosma, C., Popan, A. I., Panc, N., Filip, D., & Balc, N. (2023). Correction of Shape Error at Cut-In and Cut-Out Points in Abrasive Waterjet Cutting of Carbon Fiber Reinforced Polymer (CFRP). *Machines, 11*(8). https://doi.org/10.3390/machines11080800.

Prakash, S. (2019). Experimental investigation of surface defects in low-power CO2 laser engraving of glass fiber-reinforced polymer composite. *Polymer Composites, 40*(12), 4704–4715. https://doi.org/https://doi.org/10.1002/pc.25339.

Prasad, K., & Chakraborty, S. (2018). A decision guidance framework for non-traditional machining processes selection. *Ain Shams Engineering Journal, 9*(2), 203–214. https://doi.org/10.1016/j.asej.2015.10.013.

Rehbein, J., Wierach, P., Gries, T., & Wiedemann, M. (2017). Improved electrical conductivity of NCF-reinforced CFRP for higher damage resistance to lightning strike. *Composites Part A: Applied Science and Manufacturing, 100*, 352–360. https://doi.org/https://doi.org/10.1016/j.compositesa.2017.05.014.

Rodden, W. S. O., Kudesia, S. S., Hand, D. P., & Jones, J. D. C. (2002). A comprehensive study of the long pulse Nd:YAG laser drilling of multi-layer carbon fibre composites. *Optics Communications, 210*(3), 319–328. https://doi.org/https://doi.org/10.1016/S0030-4018(02)01807-2.

Roldan-Jimenez, L., Bañon, F., Valerga, A. P., & Fernandez-Vidal, S. R. (2022). Design and Analysis of CFRP Drilling by Electrical Discharge Machining. *Polymers, 14*(7). https://doi.org/10.3390/polym14071340.

Salama, A., Li, L., Mativenga, P., & Sabli, A. (2016). High-power picosecond laser drilling/machining of carbon fibre-reinforced polymer (CFRP) composites. *Applied Physics A*, *122*(2), 73. https://doi.org/10.1007/s00339-016-9607-8.

Salama, A., Yan, Y., Li, L., Mativenga, P., Whitehead, D., & Sabli, A. (2016). Understanding the self-limiting effect in picosecond laser single and multiple parallel pass drilling/machining of CFRP composite and mild steel. *Materials & Design*, *107*, 461–469. https://doi.org/https://doi.org/10.1016/j.matdes.2016.06.048.

Samsingh, V. R., Subbaraj, S., Malathi, K., Sundarsingh, E. F., Yogeshwari, P., Saffrine, K., & Rao, Y. V. R. (2017). Characterization of Delamination in Fiber-Reinforced Epoxy-Based PCB Laminates, Using an EBG-Enhanced Planar Microwave Sensor. *IEEE Transactions on Components, Packaging and Manufacturing Technology*, *7*(10), 1739–1746. https://doi.org/10.1109/TCPMT.2017.2737783.

Schneider, F., & Petring, D. (2018). CFRP is Demanding – Also on Laser Cutting. *Laser Technik Journal*, *15*(2), 61–65. https://doi.org/https://doi.org/10.1002/latj.201800012.

Shyha, I. S., Kuo, C.-L., & Soo, S. L. (2013). Workpiece surface integrity and productivity when cutting CFRP and GFRP composites using a CO2 laser. In *Int. J. Mechatronics and Manufacturing Systems* (Vol. 7, Issue 2).

Singh, K. K., & Thakur, R. K. (2022). Experimental analysis on carbon nanotube embedded GFRP composites during AWJM. *Materials and Manufacturing Processes*, *37*(2), 210–220. https://doi.org/10.1080/10426914.2021.1960997.

Sohaili, J., Kumari Muniyandi, S., Suhaila Mohamad, S., & Ariffin, A. (2019). Effect of particle size and amount of nonmetallic PCB materials on the mechanical properties of rHDPE/PCB composites. In */ Malaysian Journal of Fundamental and Applied Sciences* (Vol. 15, Issue 2).

Solati, A., Hamedi, M., & Safarabadi, M. (2019). Comprehensive investigation of surface quality and mechanical properties in CO2 laser drilling of GFRP composites. *The International Journal of Advanced Manufacturing Technology*, *102*(1), 791–808. https://doi.org/10.1007/s00170-018-3164-6.

Soliman, Eslam, Al-Haik, Marwan, & Taha, Mahmoud Reda. (2012). On and off-axis tension behavior of fiber reinforced polymer composites incorporating multi-walled carbon nanotubes. *Journal of Composite Materials*, *46*(14), 1661–1675. https://doi.org/10.1177/0021998311422456.

Staupendahl, G. (2014). A Novel Q-switched CO2 Laser and its Applications. *Laser Technik Journal*, *11*(5), 22–25. https://doi.org/https://doi.org/10.1002/latj.201400050.

Tao, N., Chen, G., Yu, T., Li, W., & Fan, L. (2020). Dual-beam laser drilling process for thick carbon fiber reinforced plastic composites plates. *Journal of Materials Processing Technology*, *281*, 116590. https://doi.org/https://doi.org/10.1016/j.jmatprotec.2020.116590.

Thakur, R. K., Singh, K. K., & Ramkumar, J. (2021). Impact of nanoclay filler reinforcement on CFRP composite performance during abrasive water jet machining. *Materials and Manufacturing Processes*, *36*(11), 1264–1273. https://doi.org/10.1080/10426914.2021.1906896.

Vigneshwaran, S., Uthayakumar, M., & Arumugaprabu, V. (2018). Abrasive water jet machining of fiber-reinforced composite materials. *Journal of Reinforced Plastics and Composites*, *37*(4), 230–237. https://doi.org/10.1177/0731684417740771.

Vishwakarma, R., Kesarwani, S., Verma, R. K., Debnath, K., & Davim, J. P. (2023). Using Harris Hawk algorithm for experimental study on the hole dilation mechanism during Micro-machining (μM) of Graphene nanoplatelets/Carbon fiber (GnP/C) reinforced polymeric composite. *Materials Research Express*, *10*(2). https://doi.org/10.1088/2053-1591/acbc67.

Vishwakarma, R., & Verma, R. K. (2021). Micro electric discharge machining (μ-EDM) of polymer nanocomposites modified by graphene nanoplatelets/carbon using rotating electrode tool. *Journal of Micromechanics and Microengineering*, *31*(8), 85010. https://doi.org/10.1088/1361-6439/ac11cb.

Voisey, K. T., Fouquet, S., Roy, D., & Clyne, T. W. (2006). Fibre swelling during laser drilling of carbon fibre composites. *Optics and Lasers in Engineering*, *44*(11), 1185–1197. https://doi.org/https://doi.org/10.1016/j.optlaseng.2005.10.008.

Wang, H., Yuan, R., Zhang, X., Zai, P., & Deng, J. (2023). Research Progress in Abrasive Water Jet Processing Technology. *Micromachines*, *14*(8). https://doi.org/10.3390/mi14081526.

Wang, X., Zheng, H., Wan, Y., Feng, W., & Lam, Y. C. (2018). Picosecond Laser Surface Texturing of a Stavax Steel Substrate for Wettability Control. *Engineering*, *4*(6), 816–821. https://doi.org/https://doi.org/10.1016/j.eng.2018.10.006.

Yalukova, O., & Sárady, I. (2006). Investigation of interaction mechanisms in laser drilling of thermoplastic and thermoset polymers using different wavelengths. *Composites Science and Technology*, *66*(10), 1289–1296. https://doi.org/https://doi.org/10.1016/j.compscitech.2005.11.002.

Appendix 1

S	Spindle speed
F	Feed rate
t	Thickness of sample
wt.%	Weight %
D	Depth of cut
T_h	Thrust force
T	Torque
R_a	Surface roughness
F_d	Delamination factor
FS	Flexural strength
MRR	Material removal rate
C_e	Circularity error
GO	Graphene oxide
t_m	Machining time
w_i	Initial weight
w_f	Final weight
ρ	Density
F_x	Axial force
F_y	Radial force
F_z	Torque force
D_{max}	Maximum diameter
D_0	Nominal diameter
W_{max}	Maximum slot cut
W_0	Minimum slot cut
R^2	Coefficient of determination
F_c	Cutting Force
R_z	Ten-Point Mean Roughness
Fd (IN)	Delamination Factor Inlet
Fd (OUT)	Delamination Factor Outlet
N-MRR	Normalized Material Removal Rate
N-F_c	Normalized Cutting Force
N-R_a	Normalized Surface Roughness

About the Authors

Dr. Jogendra Kumar is a research associate-I in an R&D-sponsored DST project at the Department of Mechanical Engineering, Harcourt Butler Technical University, Kanpur, India. He completed a doctorate (PhD) at the Madan Mohan Malaviya University, Gorakhpur, India, in 2021. Also, he has completed an R&D sponsored Project by the Ministry of Textiles, Government of India, as a Project Leader. His research focuses on recycling waste materials, nanomaterial synthesis and growth, nanocomposite machining, numerical modeling, and optimization. Apart from research, he worked for over 2.5 years as an assistant professor at NIMS University, Rajasthan, and Jagadambha College of Engineering and Technology, Maharashtra. He has published over 40 research articles in international peer-reviewed journals and book chapters. Additionally, he has published 10 patents/copyrights and 10 book chapters.

Professor Dr. Rajesh Kumar Verma is a professor in the Department of Mechanical Engineering, School of Engineering, Harcourt Butler Technical University (HBTU) Kanpur India, (UP Govt. Technical University). He has worked as Additional Controller of Examinations and Finance Controller (FC) of HBTU. He received his PhD (Engineering) from Jadavpur University, Kolkata, India. He is a Chartered Engineer by the Institution of Engineers (India). He is actively involved in teaching about and researching nanomaterial, polymer composites, modeling, simulation, optimization, advanced machining, and machinability estimation of composites/nanocomposite materials. Dr. Verma supervised six PhD and 13 master's theses; currently, five PhDs are ongoing. He has published over 150 peer-reviewed articles in refereed international journals and conference proceedings. The e-content and courses he developed are widely used in the Indian university system. He has more than 12 nos. Completed/Ongoing research and development (R&D) projects sponsored by various government agencies. He has 15 filed Indian patents and three India copyright registered documents as well. In 2023, Dr. Verma was the consecutive recipient of the prestigious Research Excellence Award-2023 (REA-2023) from the Honorable Vice-Chancellor, HBTU, India, and additionally, in 2024 achieved the Research Excellence Award- 2024 (REA-2024) from the H. E. Honorable Governor of Uttar Pradesh (UP) state and Chancellor HBTU Kanpur for his exemplary research work. His research proposal secured second rank in Uttar Pradesh- Pollution Control Board (UPPCB) Gomti Nagar, Lucknow in a Hackathon "Beat Plastic Pollution" on 02/10/2018, out of a total of 699 project entries and received the Best Paper Awards at the PRIME2021 and the ICMPC2021 conferences. He is a reviewer for various international publishers such as Elsevier, Sage, Springer, etc. Moreover, Dr. Verma has been invited as keynote/invited speaker, section chair, and member for several international conferences with themes in mechanical and production engineering, emerging materials, and manufacturing science. He has developed various nanomaterials for structural applications and proposed hybrid optimization modules recognized by multiple peer-reviewed journal publishers like Elsevier, Springer, Taylor and Francis, Sage, Wiley, WSPC, etc.

Dr. Yadvendra Kumar Mishra is working as an assistant professor in the Department of Mechanical and Aerospace Engineering at NIMS University in Jaipur, Rajasthan, India. He completed his doctorate (PhD) from Madan Mohan Malaviya University of Technology, Gorakhpur, in 2022 and his MTech. in materials engineering from NIT-Jaipur in 2010. He has over 15 years of teaching experience, including key educational and administrative positions in engineering colleges in the Delhi-NCR region. Dr. Mishra's research areas include laser drilling in fiber-reinforced polymer (FRP) composites, unconventional machining methods, ANSYS-based simulation, MATLAB modeling, soft computing, and multi-objective optimization strategies. He has published over 22 peer-reviewed research publications in prestigious international journals and received three patents for his innovative mechanical and aerospace engineering contributions. He has won Best Paper and Best Faculty Awards. His academic contributions include journal reviews, demonstrating his devotion to research and education.

Dr. Rajneesh Kumar Singh is a Postdoc Fellow at the Department of Mechanical Engineering at the Malaviya National Institute of Technology, Jaipur, India. He completed a doctorate (PhD) at the Madan Mohan Malaviya University, Gorakhpur, India, in 2020. His research interests focus on FEP composites, conventional machining, numerical modeling, and surface texturing. He has published over 20 articles in international peer-reviewed journals and book chapters.

Index

D

E

F

G

W

X